AF791442

Structural Condition Assessment of Existing Buildings

Other Titles of Interest

Guideline for Structural Condition Assessment of Existing Buildings (11-99), by the American Society of Civil Engineers (ASCE/SEI 2000). Standard ASCE/SEI 11-99 establishes guidelines for assessing the structural condition of existing buildings constructed of various combinations of materials. (ISBN 978-0-7844-0432-4)

Guideline for Condition Assessment of the Building Envelope (30-14), by the American Society of Civil Engineers (ASCE/SEI 2014). Standard ASCE/SEI 30-14 provides a guideline and methodology for assessing the condition and performance of existing building envelope systems and components. (ISBN 978-0-7844-1325-8)

Structural Safety and Its Quality Assurance, edited by Bruce R. Ellingwood and Jun Kanda (ASCE/SEI 2005). This report uses an international perspective to look at structural safety problems from basic concept to design and construction. (ISBN 978-0-7844-0816-2)

Residential Building Loads: Review and Roadmap for the Future, edited by Jay H. Crandell, Thomas M. Kenney, and David Rosowsky (ASCE/SEI 2006). This report identifies current knowledge and areas for additional research with regard to structural design loads for residential building applications. (ISBN 978-0-7844-0852-0)

Seismic Evaluation and Design of Petrochemical and Other Industrial Facilities, Third Edition, by the Task Committee on Seismic Evaluation and Design of Petrochemical Facilities (ASCE/Energy Division 2020). This report offers practical recommendations regarding the design and safety of new and existing petrochemical facilities during and following an earthquake. (ISBN 978-0-7844-1548-1)

ASCE Manuals and Reports on Engineering Practice No. 158

Structural Condition Assessment of Existing Buildings

A Guideline

Subcommittee on Structural Condition Assessment of Existing Buildings

Sponsored by the
Structural Condition Assessment of Existing Buildings
Subcommittee of the
Structural Condition Assessment and Rehabilitation of Buildings Committee of the
SEI Codes and Standards Activities Division of the
Structural Engineering Institute of the
American Society of Civil Engineers

Published by the American Society of Civil Engineers

Cataloging-in-Publication Data on file with the Library of Congress

LCCN 2024007119

Published by American Society of Civil Engineers
1801 Alexander Bell Drive
Reston, Virginia 20191-4382
www.asce.org/bookstore | ascelibrary.org

Any statements expressed in these materials are those of the individual authors and do not necessarily represent the views of ASCE, which takes no responsibility for any statement made herein. No reference made in this publication to any specific method, product, process, or service constitutes or implies an endorsement, recommendation, or warranty thereof by ASCE. The materials are for general information only and do not represent a standard of ASCE, nor are they intended as a reference in purchase specifications, contracts, regulations, statutes, or any other legal document. ASCE makes no representation or warranty of any kind, whether express or implied, concerning the accuracy, completeness, suitability, or utility of any information, apparatus, product, or process discussed in this publication, and assumes no liability therefor. The information contained in these materials should not be used without first securing competent advice with respect to its suitability for any general or specific application. Anyone utilizing such information assumes all liability arising from such use, including but not limited to infringement of any patent or patents.

ASCE and American Society of Civil Engineers—Registered in US Patent and Trademark Office.

Photocopies and permissions. Permission to photocopy or reproduce material from ASCE publications can be requested by sending an email to permissions@asce.org or by locating a title in the ASCE Library (https://ascelibrary.org) and using the "Permissions" link.

***Errata:* Errata, if any, can be found at https://doi.org/10.1061/9780784485422.**

Copyright © 2024 by the American Society of Civil Engineers.
All Rights Reserved.

ISBN 978-0-7844-1621-1 (soft cover)
ISBN 978-0-7844-8542-2 (PDF)
ISBN 978-7844-8543-9 (ePub)

Manufactured in the United States of America.
28 27 26 25 24 1 2 3 4 5

MANUALS AND REPORTS ON ENGINEERING PRACTICE

(As developed by the ASCE Technical Procedures Committee, July 1930, and revised March 1935, February 1962, and April 1982)

A manual or report in this series consists of an orderly presentation of facts on a particular subject, supplemented by an analysis of limitations and applications of these facts. It contains information useful to the average engineer in his or her everyday work, rather than findings that may be useful only occasionally or rarely. It is not in any sense a "standard," however, nor is it so elementary or so conclusive as to provide a "rule of thumb" for nonengineers.

Furthermore, material in this series, in distinction from a paper (which expresses only one person's observations or opinions), is the work of a committee or group selected to assemble and express information on a specific topic. As often as practicable, the committee is under the direction of one or more of the Technical Divisions and Councils, and the product evolved has been subjected to review by the Executive Committee of the Division or Council. As a step in the process of this review, proposed manuscripts are often brought before the members of the Technical Divisions and Councils for comment, which may serve as the basis for improvement. When published, each manual shows the names of the committees by which it was compiled and indicates clearly the several processes through which it has passed in review, so that its merit may be definitely understood.

In February 1962 (and revised in April 1982), the Board of Direction voted to establish a series titled "Manuals and Reports on Engineering Practice," to include the manuals published and authorized to date, future Manuals of Professional Practice, and Reports on Engineering Practice. All such manual or report material of the Society would have been refereed in a manner approved by the Board Committee on Publications and would be bound, with applicable discussion, in books similar to past manuals. Numbering would be consecutive and would be a continuation of present manual numbers. In some cases of joint committee reports, bypassing of journal publications may be authorized.

A list of available Manuals of Practice can be found at https://ascelibrary.org/page/books/s-mop.

DEDICATION

All the authors involved with this Manual of Practice would like to dedicate this edition to the memory of Carl A. Baumert Jr., who passed away in April 2018. We would like to thank Carl for his dedication and commitment to the revision of this publication and for his tutelage of its committee members throughout the years. Carl was the past chairman of the main and both structural assessment subcommittees. During his chairmanship, the second edition of the ASCE Standard Guideline for Structural Condition Assessment of Existing Buildings was published, and the revisions to the third edition were commenced. Carl's creative contributions regarding historic structures and materials are present throughout this book.

It is with great pleasure that we dedicate this Manual of Practice to the memory of Carl A. Baumert Jr.

CONTENTS

STRUCTURAL CONDITION ASSESSMENT OF EXISTING BUILDINGS SUBCOMMITTEE

The Subcommittee on Structural Condition Assessment of Existing Buildings of the Structural Condition Assessment and Rehabilitation of Buildings Committee of the SEI Codes and Standards Activities Division of ASCE's Structural Engineering Institute prepared this manual. The Task Committee Member authors were

George R. Mulholland, P.E. (*Chair*)
Raths, Raths & Johnson, Inc.

William A. Andrews
Walter P. Moore

James A. D'Alosio
Forensic & Failure Analysis, Inc.

David A. Deress
Wiss, Janey, Elstner Associates, Inc. (Retired)

John E. Durnsick
U.S. Department of State

Dan Eschenasy
Gilsanz Murray Steficek

Robert C. Field
South Street Structural Engineers

Melvyn Green, P.E., R.Eng, S.E.
Melvin Green & Associates, Inc.

Ying Huang
North Dakota State University

Camile A. Issa
Lebanese American University

David P. Martin
Foley Buhl Roberts & Associates, Inc.

Paul L. Millman
Superstructures

Antonio Nanni
University of Miami

Joseph F. Neussendorfer
U.S. Construction Research

Lance Parker
Thornton-Tomasetti/LZA Tech

Gregory P. Parsons
Builders Engineering Corp.

Robert T. Ratay
Ratay Structures

Joseph Riddle
Engineering Systems, Inc.

Christina E. Salchow
Thornton-Tomasetti

Chad M. Schrand
Horner & Shifrin

Mohammadreza Sharafbayani
Bechtel OG&C

Jeffery A. Travis
Exponent Failure Analysis Associates

Brian E. Trimble, P.E.
International Masonry Institute

Additional contributions were made by ASCE members who were not on the subcommittee, and their names are listed subsequently.

Ronald W. Anthony
Anthony-Associates, Inc.

Gregory T. Holbrook, P.E.
McKim & Creed, Inc.

BLUE-RIBBON PANEL REVIEWERS

We thank the following members of the Blue Ribbon Panel for their careful review of the document and excellent recommendations for improvements.

Kevin Conroy, P.E., S.E.
Simpson, Gumpertz & Hager, Inc.

Thomas Smith, AIA, RRC
T.L. Smith Consulting, Inc.

Tore Arnesen, P.E., S.E.
Vector Corporation (Retired)

Heather Anesta, P.E., S.E.
Anesta Consulting, Inc.

Randall Bernhardt, P.E., S.E.
Bernhardt Forensic Engineering

Ayman Sharma, Ph.D., P.E.
AS Consulting Engineering Services

PREFACE

Changing economic conditions, concern for historic preservation, emphasis on fully utilizing conveniently located structures, space shortages, and increasing cost of materials and products used in the construction of new buildings have resulted in a need to evaluate and more fully utilize the existing building inventory. In addition, because replacing existing buildings with new construction releases additional greenhouse gases and heat energy into the environment, there are environmental benefits to reusing rather than replacing existing structures. Particularly in older cities, emphasis has shifted from replacement to preservation, rehabilitation, and strengthening of existing buildings. Equally important is the need to evaluate the existing structure relative to its remaining service life, which must also be considered to ensure that an existing building assessment addresses the overall health of the building, including but not limited to initial design parameters, lateral and vertical load path continuity, construction quality, age, material deterioration, repairability, cycle of repairs, and future use parameters. Such considerations are independent of owner, client, and municipality interests.

New processes resulting in changes of building systems and business equipment frequently impose greater loads on an existing building structure and may also require additional openings and restructuring. More stringent building code provisions for design load requirements, improved seismic resistance, accessibility, or fire protection may also demand retrofitting or structural reinforcement. Also, any known site conditions should be reviewed to determine whether modifications to structural systems are required or even possible.

Adaptive reuse, rehabilitation, and improvement of existing buildings all require a thorough assessment of present building performance and capability for use by owners, designers, building officials, and contractors. Current data for assessment are dispersed and not readily available to many of those making technical decisions. Such information has been compiled and subjected to a consensus review by this committee. The intent of this Manual of Practice is to provide the design community with suggested guidelines on building condition assessments for selected materials and for other areas related to the structural performance of buildings. To that end, this publication has been prepared for use by qualified professional engineers and regulatory officials.

This publication has been adapted from ASCE 11-99, *Guideline for Structural Condition Assessment of Existing Buildings*. This reimagined adaptation, as a Manual of Practice, has been reformatted to make this manual more user-friendly. The major change has been to organize the document by material type and not by test or assessment type. We hope this change will make the use of this document more efficient. This publication also serves as an updated resource to be utilized prior to any future revisions of ASCE 11-99.

These suggested guidelines are not intended to be all-inclusive or prescriptive. Methods and procedures are presented as a reference resource. Other rational methods and procedures are not only permissible but also encouraged so long as they are deemed reliable by the authority having jurisdiction and sufficient comparisons are available with recognized methods.

Inasmuch as interpretation of the results of the evaluation must be based on the professional experience and judgment of the practitioner, it is not addressed as a part of this Manual of Practice.

People performing the evaluation may be exposed to hazardous materials, operations, and equipment during structural condition assessments. This manual does not purport to address all safety issues associated with its use. It is the responsibility of the manual user to establish appropriate safety and health practices and determine the applicability of regulatory limitations prior to use.

The material presented in this publication has been prepared in accordance with recognized engineering principles. This manual should not be used without first securing the services of a qualified design professional to provide advice with respect to its suitability for any given application. The publication of the material contained herein is not intended as a representation or warranty on the part of ASCE or of any other person named herein that this information is suitable for any general or particular use or promises freedom from infringement of any patent or patents. Anyone making use of this information assumes all liability for such use.

ACKNOWLEDGMENTS

These acknowledgments are made to honor the contributions of the following individuals to the creation of this Manual of Practice:

William Andrews
Heather Anesta
Ronald W. Anthony
Tore Arnesen
Randall Bernhardt
Kevin Conroy
James D'Alosio
David Deress
John Durnsick
Dan Eschenasy
Robert Field III
Melvyn Green
Gregory T. Holbrook
Ying Huang
Camille Issa
David Martin
Paul Millman
George Mulholland
Antonio Nanni
Joseph Neussendorfer
Lance Parker
Gregory Parsons
Robert Ratay
Joseph Riddle
Guillermo Riveros
Christina Salchow
Chad Schrand
Ayman Shama
Mohammadreza Sharafbayani
Thomas Smith
Jeffrey Travis
Brian Trimble

CHAPTER 1
GENERAL

1.1 SCOPE AND INTENT OF THE MANUAL OF PRACTICE

The intent of this publication is to provide guidelines and methodology for assessing the structural condition of existing buildings constructed of combinations of materials such as concrete, metals, masonry, and wood. This guideline establishes the assessment procedure, including investigation, testing methods, and format for the report of the condition assessment.

This book is intended as a guide to the design professional in providing comprehensive information for clients such as building owners, prospective purchasers, tenants, regulatory officials, and others. Any evaluation will involve engineering judgment and contain factors that cannot be readily defined and standardized. A section providing guidance for evaluations is also included. This section should be used by the design professional as part of their structural evaluation. The Local Existing Building Code and/or the International Existing Building Code (IEBC—currently adopted edition) should be consulted for code requirements regarding the evaluation and repairs of existing buildings.

This publication does not specifically address seismic retrofitting of structures. The investigator should reference ASCE 31, *Seismic Evaluation of Existing Structures,* and ASCE 41, *Seismic Rehabilitation of Existing Structures,* for seismic evaluation and rehabilitation of structures, respectively.

Dimensions and quantities in this book are expressed in inch-pound Imperial (customary) units followed by conversion to SI units in parentheses.

1.2 PURPOSE OF ASSESSMENT

A structural condition assessment of an existing building can be undertaken for a number of purposes. These can include developing a performance report, establishing building use, serviceability, code compliance, life safety, durability, historic preservation, determining remaining service life, or a number of special purposes on the basis of the specific building and its current or proposed occupancy or function. The design professional should consider the possibility of the presence of hazardous materials or conditions when assessing an existing building, advise the client as necessary, and take or recommend appropriate precautions. The design professional should comply with local ordinances that may require governmental entities to be notified of life safety issues.

Structural condition assessment purposes vary on the type of structure and the needs of the client or owner. The purpose should be clearly defined and communicated to the client or owner. The range of purposes includes, but are not limited to,

- Identifying any visually discernible conditions of damage, distress, or deterioration. This may be accomplished with a limited walk-through visual review.
- Increasing the level of confidence of the future performance of the structure. This typically entails a more comprehensive site reconnaissance, including removal of finishes or toppings at specific areas to observe the structural elements, performing intrusive or destructive investigation work, and possibly material testing, mathematical analyses, and/or load testing.
- Determining or estimating the remaining service life of the structure. This may be accomplished by considering the loads and intended use of the structure, assessing the condition of the materials, confirming the continuity of the vertical and lateral load paths, reviewing the feasibility of repairs and cycle of repairs, and evaluating the ability for the structure to withstand current and future loads.

1.2.1 Condition Assessment Scope Level

Similar to the purpose of an assessment, scope levels may vary between projects and should be well-documented to ensure that the expectations of the client are met. Conceptually, these levels include the following.

1.2.1.1 Preliminary Assessment. This consists of a brief walk-through of a structure or facility to identify visually discernible conditions of distress, damage, deterioration, or instability. It can be used as a stand-alone review protocol, or it can be performed to help better define a more comprehensive assessment. It includes document review, a limited site reconnaissance, and a verbal or written report.

1.2.1.2 Detailed Assessment. The extent of a more comprehensive assessment varies. It may or may not include intrusive or destructive investigations, material sampling and testing, mathematical analysis, and/or load testing.

1.3 QUALIFICATIONS AND EQUIPMENT

1.3.1 Personnel Qualifications

All personnel involved in the assessment should possess the necessary technical qualifications, including practical experience, education, and professional judgment required to perform the individual technical tasks assigned. Interpretation of observations and test and probe results and conclusions shall be performed by a licensed design professional qualified in the appropriate discipline.

1.3.2 Equipment

Equipment should be utilized as appropriate to accomplish or perform the various tests and investigative methods specified in the manual. All equipment should be in good working order. For equipment that can be calibrated, reports of calibration shall be made available.

1.4 AGREEMENTS

1.4.1 Services

The scope of services for the structural condition assessment should be defined by the licensed design professional qualified in the appropriate discipline, and all conditions and services shall be mutually agreed upon by the client and design professional. Services may include one or more of the identified purposes of the assessment and may involve more than one specific building component or system covered in the subsections of this book. Evaluation and acceptance criteria should be defined as part of the agreement, as well as an agreed-on time frame for the review and the report. Limits or exclusions to the assessment should also be identified. Clients should be advised of the possibility of previously undiscovered deficiencies and hazardous materials being present in an existing building, even if a previous investigation has been completed. Finally, the deliverables should be defined; usually, a report stamped and signed by a licensed design professional, possibly including representative photographs and any sketches, test results, analysis results, and so on.

1.4.2 Compensation

The client and the qualified design professional should mutually agree upon the design professional's compensation for the services specified in Section 1.4.1. The agreement shall specify whether it includes reimbursable expenses such as travel costs, testing laboratory costs, or other expenses that are not labor-related. Compensation provisions shall consider, after a review of the preliminary data, that additional investigations may be recommended and, if accepted, change the scope of the project.

1.4.3 Authority and Accessibility

The agreement shall clearly identify the scope and authority of the design professional to perform the necessary investigation and tests and shall assure access to the site, the building, the various portions of the building requiring investigation, and drawings and documents available to perform the condition assessment. Responsibility for access to (e.g., scaffolding, manlift, swing stage), removal of, and repairs to finishes and building elements that may result from the evaluation, tests, or investigative procedures should be identified.

1.4.4 Liability

The extent of liability, if any, expressly accepted by the design professional shall be stated in the agreement. Requirements for liability coverage shall be clearly documented.

1.5 DEFINITIONS

The following definitions are provided to assure a uniform understanding of some selected terms as they are used in this manual of practice:

1. **Assessment**: Systematic collection and analysis of data, documentation, evaluation, and recommendations regarding the various portions of an existing building that are the subject of the investigation.
2. **Attributes**: Building performance features and characteristics such as fire safety, building behavior under loading conditions, and so on.

3. **Building**: Enclosed or partially enclosed structure for occupancy by personnel, animals, material, or equipment; any structure used or intended for supporting or sheltering any use or occupancy.
4. **Client**: Person or persons for whom the structural condition assessment of the building is being performed. A government agency, corporation, association, or other group could be the client.
5. **Component (Nonstructural)**: Portion of the building able to transfer applied lateral and/or vertical loads and its own weight to structural components. This portion may include ceilings, stairs, parapets, ornamentation, cornices, overhangs, equipment, flagpoles, chimneys, infill walls, and handrails.
6. **Component (Structural)**: Portion of a building designed to resist applied lateral and/or vertical loads, including those imposed by other structural or nonstructural components and its own weight. These components include, for example, roof decks, floors, walls, elevator cores, stair shafts, frames, foundation structures, piles, and piers.
7. **Computational Analysis**: Quantitative evaluation of a building component or system by or under the direction of a design professional.
8. **Connection**: Fastening between components or other parts of the building, which, either by design or as a result of construction, is capable of transferring loads or forces. A "connector" is a mechanical connecting device.
9. **Design Professional**: Architect or engineer licensed or registered to practice in the governmental jurisdiction in which the building is located.
10. **Destructive Testing**: Testing properties of materials, components, or systems to a point at and beyond which the element, component, or system can no longer fulfill its function. Destructive testing of portions of a building may be carried out on the building in situ or on samples removed and tested in the laboratory.
11. **Evaluation**: Process of assessing the structural adequacy of the building or component for its intended use and/or performance. Evaluation, by its nature, implies the use of personal and professional judgment by those functioning in the capacity of design professionals.
12. **Investigation**: Activity of examining, measuring, testing, gauging, and using other procedures to ascertain the quality or state, detect errors, defects, or deterioration, and otherwise appraise materials, components, systems, or environments.
13. **Nondestructive Testing**: Measured assessment of a material, component, or system without altering their properties or impairing future service.

14. **On-Site (In Situ) Load Testing**: Testing carried out on building systems or components on-site to determine structural performance characteristics. It may include artificially applied loads intended to replicate service or ultimate load conditions.
15. **Professional Architect or Engineer**: *See* Design Professional.
16. **System**: Assemblage of structural components within a building that, by contiguous interconnection, form a path by which external and internal forces applied to the building are delivered to the ground.
17. **Testing Laboratory**: Facility in which the characteristics or performance of materials and products are measured, examined, tested, calibrated, or otherwise determined. Depending on the test procedure, some testing agencies and technicians must be accredited or certified.

1.6 REFERENCES

References are significant to the use of this manual. The more the user makes use of these references and becomes familiar with them, the more valuable the book will be. It should be noted that many references are taken from the field of bridge engineering, even though the subject of this manual is buildings. Much of the research on material problems, tests on samples of the materials, and nondestructive testing has been in the bridge engineering discipline. Bridge structural systems are environmentally exposed and, therefore, subject to more severe weathering, have less redundancy, and typically cannot be abandoned as they deteriorate. These references are valuable because the same materials are used in both bridges and buildings.

Specific references are located within Appendixes B1 through B4, while organizational references are located in Appendix B5.

CHAPTER 2
ASSESSMENT PROCEDURE

2.1 GENERAL

The process of assessing the structural condition of an existing building consists of assembling and systematically analyzing information and data regarding the building or portions thereof to determine the structural adequacy. A general procedure for conducting a structural assessment and evaluation of existing buildings is shown in Figure 2-1 and described herein. The procedure is not necessarily complete, and the ordering of the steps in the procedure depends on the specific purpose of the investigation, the availability of design and construction documents, and the number of buildings encompassed in the investigation. The precise procedure for the condition assessment of each individual building should be determined by the design professional in charge of the investigation. This guideline does not preclude the use of other procedures or address the rapid evaluation method of a structure that has been exposed to an extreme event. Those in charge of evaluating these types of structures to determine whether they are stable and safe for building occupancy, first responders, or emergency personnel should reference ATC 20-1, *Post-Earthquake Safety Evaluation of Buildings* (ATC 2005), and ATC-45, *Field Guide: Safety Evaluation of Buildings after Windstorms and Floods* (ATC 2004).

2.2 APPROACH

Because of the potential cost of a comprehensive structural assessment for an existing building, a multilevel approach is usually recommended. The basic process typically entails a preliminary assessment followed by a

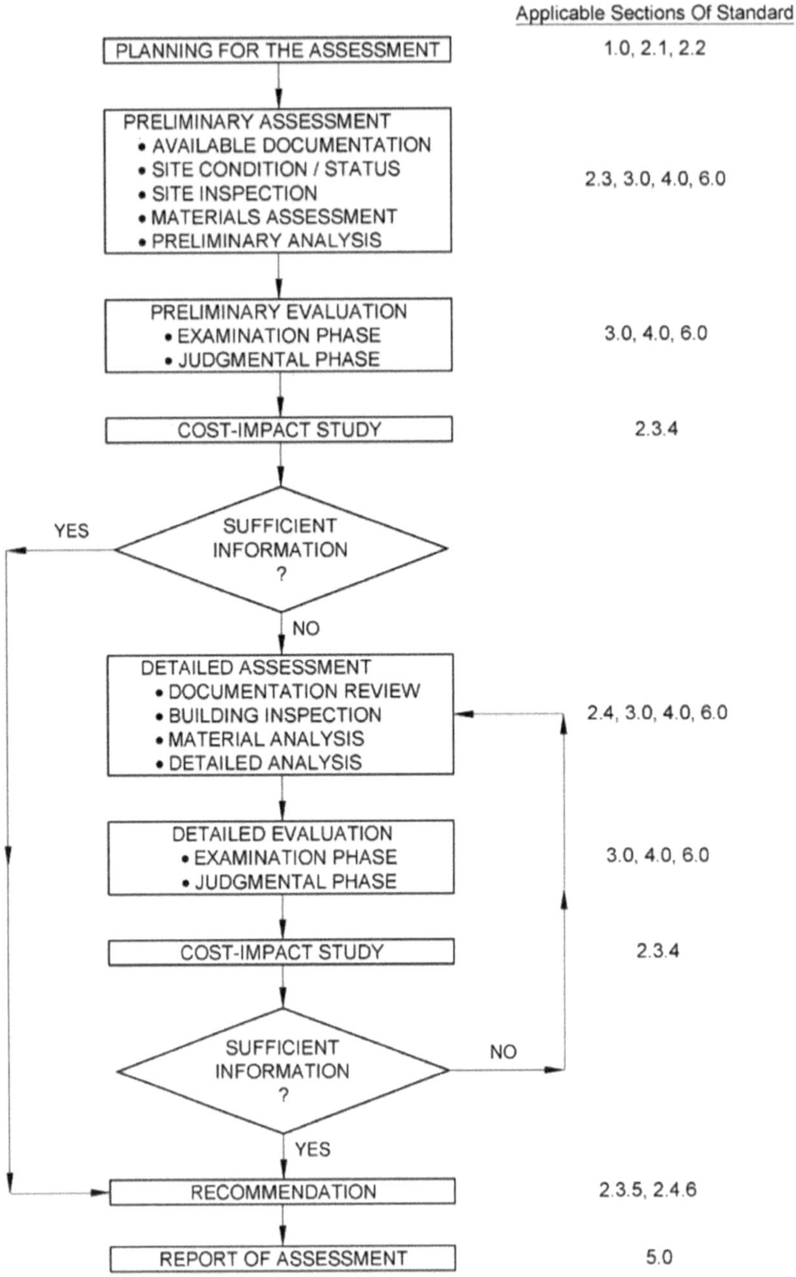

Figure 2-1. General structural assessment and evaluation procedure for existing buildings.
Source: Courtesy of Raths, Raths & Johnson, Inc.

detailed assessment. If deemed necessary or beneficial, a cursory assessment (Chapter 5, Section 5.2.2.1) could be performed to devise a strategy for the more detailed building assessment. When it has already been determined that the building requires a detailed assessment, cursory and preliminary assessments may be eliminated. If the scope of the investigation encompasses a large inventory of buildings, the design professional should establish screening criteria to select buildings that will be included in the preliminary assessment. Potential screening factors, such as date of construction, building area, acquisition cost, importance, and occupancy or use, can be used to reduce and prioritize the inventory for the preliminary assessment. The purpose of inventory reduction is to minimize or eliminate unnecessary detailed investigations and to keep the scope of work within reasonable limits.

2.2.1 Condition Assessment Considerations

Several items must be considered by a design professional when performing an assessment of a structure. These items will vary from project to project. The intent of the following suggested aspects to consider is not to be regarded as a checklist but to provide useful questions to consider before and during a building assessment. The questions listed subsequently are intended to assist the design professional in understanding potential structural limitations that may exist in a building for the purpose of increasing the level of confidence in the building's future structural performance.

2.2.1.1 Physical Condition

- Is there any visually discernible damage, deterioration, or defect that may compromise structural integrity?
- Is there any visually discernible distress or deformation that may indicate the onset of manifestation of a structural limit state?
- Is there any visually discernible damage or loss of continuity to the lateral and vertical load path components?

2.2.1.2 Environmental Exposure

- Has the structure been subjected to, or will it be exposed in the future to, environmental conditions that can cause deterioration, such as settlement, moisture, corrosive materials, or high heat?

2.2.1.3 Known Cautionary Conditions

- Are there existing members or structural systems with known limitations? (i.e., pre-1975 glue-laminated beams, bowstring trusses)

- Are there unbraced conditions such as supports at cantilever beam lines, long compressive beam flanges, or long truss compression members?

2.2.1.4 Load Path

- Does the structure have well-defined load paths for gravity, lateral, and applied loads?
- Are the gravity and lateral load path components visible and able to be inspected? Are they well-defined and in good condition?

2.2.1.5 Loading

- Has the structure been subjected to loading or loads in excess of that for which it was likely designed or to extraordinary loads such as fire?
- Is the structure being repurposed such that the code-mandated loads will be greater than the original design requirements?

2.2.1.6 Occupancy

- Is the building currently, or will it be in the future, a high-importance essential facility?
- Will it be used for public occupancy?
- Is it a temporary or seldom-used structure?
- Has or will the building experience a change in occupancy from one IBC category to another?

2.2.1.7 Design Quality Confidence

- Is there a high degree of confidence that the initial, renovation, and/or repair designs were performed competently?
- If not, are the structural elements visible and measurable?

2.2.1.8 Construction Quality Confidence

- Is there a high degree of confidence that the initial, renovation, and/or repair construction were performed in accordance with the design documents?
- If not, is the level of construction quality well-documented?

2.2.1.9 Concealed Defect Potential

- Does the type of construction have the possibility for concealed defects that may significantly limit the performance of the structure?

2.2.1.10 Potential Failure States

- Is the structural system known to be subject to failure mechanisms that can result in sudden failure without warning signs?
- If so, are there any secondary support systems?
- Could the failure of the considered system result in a more widespread or progressive collapse?

2.2.1.11 Significance of Considered Assembly

- Is the considered structural system or mechanism a local element, or does it involve a primary structural element of the building?
- Could the failure of the considered system result in danger to the public and occupants?

2.2.1.12 Unusual Conditions

- Are there structural conditions that are unusual or disproportionate (significant spans, load collection elements, or other conditions) that just do not "seem" right?

2.3 PRELIMINARY ASSESSMENT

The preliminary assessment provides the initial analytical data for estimating the structural adequacy of an existing building and for establishing the need and priority for a more detailed analysis. The basic steps include (1) available documents review, (2) site investigation, (3) preliminary analysis, and (4) preliminary findings and recommendations.

2.3.1 Available Documents Review

For the preliminary assessment, design information such as drawings and specifications, design criteria, soil investigations, calculations, and other information will, if available, permit an easier and more accurate evaluation and expedite the site investigation. When possible, such information should be assembled prior to the site investigation. When such information is not available, the design professional must obtain the required information and data during the site investigation.

2.3.2 Structural Site Investigation

Structural site investigations enable the design professional to better confirm the accuracy of the existing design information and to assess the structural condition of the building. Any evidence of structural modifications, deterioration of materials, discrepancies in documentation, overstress in structural members or connections, settlement or foundation problems, or unique structural features should be noted. The investigation team should include at least one design professional experienced in structural evaluation. Also, personnel familiar with the building, such as the building supervisor, should accompany the investigation team to provide access to various areas within the building, describe functional requirements, and point out any known areas of modification, deterioration, or damage.

The type and levels of site investigations vary:

- Visual: In structures such as buildings, much of the structural elements are concealed by finishes. Some, but not all, types of building distress may make themselves visually discernible in the finishes.
- Destructive/Nondestructive: This may be especially appropriate for structures for which no reliable information is available about the design or the construction quality of the structure. There are two basic types: (1) Intrusive destructive—This includes removal of finishes at specific locations to observe the structural elements, as well as test probes into the structural materials, such as reinforced concrete and masonry elements. (2) Nondestructive—This includes radar and X-ray testing. It also includes sounding of concrete surfaces to detect interior delaminations that may not present themselves visually.

2.3.3 Preliminary Analysis

The preliminary analysis provides the basis for (1) estimating whether the building has adequate capacity to withstand the specified structural criteria, which may include code compliance; (2) assessing property damage; and (3) identifying structural deficiencies in the building with special reference to applicable minimum life and public safety standards and codes. Engineering judgment and experience, in conjunction with simplified analysis techniques, are required to determine rational structural demands and capacities for the critical members of the building. This analysis may indicate the need for a more rigorous analysis. Environmental effects on the structure should be considered as part of the assessment. The following subsections provide a discussion of key aspects of the preliminary analysis.

2.3.3.1 Loading and Performance Criteria. The design professional should establish and review the loading and performance criteria to be used in the preliminary analysis, such as the applicable current national, state, or local building codes and standards; federal regulations; or owner-prescribed requirements. Moreover, the design professional should become familiar with the design code and detailing practices in effect when the building was constructed. The performance criteria should establish the essential performance requirements imposed on the building, such as basic life safety of the occupants or operational capability, whether the building has experienced a change of occupancy, additions, repairs, renovations that could change load paths, importance categories, and the need for upgrades.

2.3.3.2 Structural Components. On the basis of a review of the available documentation, the results of the on-site investigation, and the loading and performance criteria, the design professional should identify the primary vertical and lateral load resisting systems in the building and the primary vertical and lateral load paths that transfer the forces to the foundation. Structural members and connections, including those members deemed critical, in the vertical, transverse, and longitudinal directions of structural systems should be identified, and the physical properties and details for these members and connections must be determined.

2.3.3.3 Material Properties. The existing design documents may contain information regarding the material properties used in the design. If not available, the strength of the materials can be estimated on the basis of the design criteria and the types of structural materials commonly used at the time of design and construction. An estimate of the material properties is typically adequate for the preliminary assessment. If the preliminary assessment reveals that certain material properties have a significant impact on the building's structural adequacy, in situ or lab testing may be justified. Both nondestructive and destructive procedures may be necessary to establish more realistic properties for the detailed assessment. Chapter 6 provides a summary of the physical properties of various construction materials and tests that can be conducted to determine the physical properties.

2.3.3.4 Component Analysis. The critical member(s) and connections in the structural system should be analyzed to determine their resistant capacities for moment, torsional, axial, and shear forces and to compare their resistant capacities to demands associated with the specified loading and performance criteria. Strength reduction factors may be used to account for uncertainties, such as detailing practices that are different from those currently recommended, the condition of the structure and material, and the section properties of the member.

2.3.3.5 Nonstructural Components. If the existing building has nonstructural components that may significantly contribute to its structural resistance, the components should be included in the preliminary analysis. The interaction of the nonstructural components with the structural framing systems or the anchorage and attachment of these nonstructural components must be considered by the design professional. The integrity of nonstructural components may also be an important consideration.

2.3.3.6 Structural Evaluation. The design professional should integrate the information and data regarding the existing building with the results from the preliminary evaluations of critical components and connections to determine the overall structural condition of the building. Chapter 4 provides information on the evaluation process, and Chapter 6 provides information on the evaluation of various construction materials. If the structural evaluation identifies a marginal deficiency, a more detailed evaluation may be required. If the structural evaluation indicates significant deficiencies, a rehabilitation study is warranted, or a recommendation can be made to phase the building out of use or prohibit occupancy until remediation.

2.3.4 Cost–Impact Study

A cost–impact study may be conducted for a building to be assessed. The cost of structural rehabilitation is subject to many factors; however, the cost for certain types of structural strengthening work can often be roughly estimated. Such an estimate can form the basis for an initial decision regarding the economic feasibility of the rehabilitation project. If the results of this initial evaluation indicate that strengthening is economically feasible, a more detailed analysis may be required. A detailed analysis may provide a reasonably accurate estimate of the total costs and determine the economic feasibility of the overall project, including such factors as disruption costs associated with temporarily vacating the building. If the cost of rehabilitation is determined to be prohibitive, alternate uses of the building may be established, or a plan made for phasing out its use.

2.3.5 Preliminary Condition Evaluation and Recommendations

The results of the preliminary assessment shall be summarized in a report (Chapter 5). The report should include, but not be limited to, the following items: (1) description of loading and performance criteria considered; (2) description of the building; (3) description of the preliminary evaluation process; (4) discussion of the preliminary findings; and (5) recommendations concerning the need for particular actions, such as conducting a detailed assessment.

2.4 DETAILED ASSESSMENT

A detailed assessment may be performed on the building being analyzed as a result of the findings of the preliminary assessment or as directed by the authority having jurisdiction or by the client. The primary purposes of the detailed assessment are as follows: (1) to determine if the building satisfies the required performance criteria or if it requires rehabilitation, and (2) if the building requires rehabilitation to identify its deficiencies and recommend alternatives for rehabilitation. The detailed assessment process is similar to that used in the preliminary assessment, except that it is done in greater detail and with more accuracy to increase the reliability of the resulting recommendations.

2.4.1 Document Review

Documents governing or prepared for the original design and construction, rehabilitation, alteration, repair of, or addition to the building should be reviewed, if available. This documentation may include drawings, specifications, material test reports, calculations, geotechnical investigations and foundation reports, standards, and codes in effect at the time of construction, construction records, operations, and maintenance data, records of alterations made to the building, information on original materials and components in the building, and site records. Possible sources of available building documentation include (1) building owner and on-site staff interviews; (2) local building department records; (3) interviews with design professionals (architects and engineers) and contractors associated with the original construction or subsequent modification of the structure (or their successor firms); (4) files of the original or successor design firms and contractors; (5) standards and local codes in effect at the time of construction; (6) files of insurance companies that may maintain drawings or records of their insured buildings; (7) professional publications dated within approximately one year of the building's design and/or construction or subsequent modification(s); (8) local library, historical society, and newspaper files; (9) manufacturer data on materials, components, and systems; (10) weather data; and (11) topographical information.

2.4.1.1 Building Features. Available design drawings, and calculations, if appropriate to the investigative scope, should be carefully reviewed to identify irregularities in the building's configuration or other features that will have a major influence on its behavior. Irregularities in the plan and vertical configuration may significantly affect the behavior of a building under certain loading conditions, and the design professional should carefully assess the impact of such irregularities. Common irregularities include mass and resistance eccentricity; common geometric irregularities, such as +, H, U,

E, T, and L – shapes; unsymmetrical geometry with respect to the vertical axis of the building; setbacks at one or more levels; discontinuities in diaphragm stiffness; building expansion joints; and discontinuous shear walls. The support of large equipment loads by the structure should also be considered.

2.4.1.2 Structural System. The type and continuity of the existing structural system have a major effect on the performance of the building. Flexible buildings can undergo large deformations and relative movements; therefore, potential damage to nonstructural elements and systems is greater. Stiff buildings will undergo smaller deformations, and the potential damage to nonstructural elements and systems is less. Rigid diaphragms tend to reduce the effects of unsymmetrical form. The most important factor, however, is the degree to which the building is tied together as a unit, with all connections having adequate capacity for load transfer by shear, tension, compression, or moment as applicable. The design professional should carefully evaluate the integrity of the structural system. This evaluation should include an assessment of the redundancies and discontinuities in the building structural system, the structural integrity of connections that tie the building together, the toughness and ductility of the construction material, the regularity of the building both in plan and in vertical configurations, load paths, and the effects of unusual structural features. Most of the factors cited are difficult to quantify; therefore, it is essential that an experienced design professional perform the structural assessment. Common problems include discontinuities with the vertical- and lateral load–resisting systems, lack of ties and continuity, lack of connector elements, and nonredundant systems.

2.4.2 Building Investigation

After the design and construction documents and the results of any previous investigations, evaluations, or testing reports have been reviewed, the design professional and any required consultants should inspect the building to verify that the existing structural system and components meet the intent of the design and determine whether there are any significant variations. The on-site investigation also affords an opportunity to inspect the nonstructural components, which may be hazardous or present specific problems affecting the structural system. Testing or special inspections should be made when there is insufficient data to make this determination or when as-built conditions are suspect.

2.4.3 Detailed Assessment

Although the preliminary analysis focused on critical members and connections, the detailed assessment will, in general, encompass the structural systems and their interactions as a total structural system.

2.4.3.1 Loading and Performance Criteria. The specified loading and performance criteria applicable to the detailed assessment should be reviewed for completeness. Acceptable performance criteria should be established for each potential mode of behavior, such as elastic and postyield levels, and should include capacity reduction factors, drift and deflection limitations, and damping values. The performance criteria should be used to establish the predominant performance requirement imposed on the building, such as the basic life safety of the occupants or essential operational capability.

2.4.3.2 Critical Systems/Members. Critical structural components, including members and connections comprising the main force-resisting system, should be identified on the basis of the document review, the building investigation, the results of previous investigations, and evaluation and tests, if applicable. Components may also be considered critical if their current condition is adversely affected by corrosion, settlement, deterioration, and so on.

2.4.3.3 Material Properties and Detailing Practices. A testing program may be established to find those material properties and detailing practices that do not conform to currently accepted practices. Test results may make the difference between accepting a building in an as-built condition or requiring a costly modification. Structural capacities of existing materials should be determined in accordance with the criteria and testing requirements in Chapter 6.

2.4.3.4 Capacities of Existing Systems/Members. The bending moment, shear, torsional shear, and axial load capacities of the critical structural systems, including members and connections comprising such systems, should be determined using the code-enforced evaluation method (ASD or LRFD) at the time of the construction. The material and section properties data for an existing building should be interpreted carefully because the detailing may not conform to new construction requirements. Interaction curves may be required to establish the capacity of columns and walls.

2.4.3.5 Required Capacities of Systems/Members. The specified vertical and lateral loading criteria should be used to determine the required capacities of the critical structural systems, including members and connections, which must resist the specified loads and satisfy the drift and deflection limitations.

2.4.3.6 Nonstructural Components. If appropriate, the nonstructural components should be evaluated to ensure that they resist the prescribed forces and deformations. The effect of nonstructural components on the building's performance should also be considered. Known deficiencies should be noted.

2.4.3.7 Structural Evaluation. The building should be evaluated by means of an actual/required capacity comparison. If the building meets the performance criteria and does not have any deficiencies, the structural condition is adequate. If it does not conform to the performance criteria, a more sophisticated analysis procedure to evaluate the building may be used. If the building still does not conform to the acceptable performance criteria, a recommendation to provide the required performance identified should be formulated.

2.4.4 Detailed Findings

The detailed findings should indicate whether the existing building conforms to the performance criteria and should state the basis for this finding. If the building does not conform to the performance criteria, the deficiencies should be clearly identified and explained. Alternate concepts for correcting the deficiency or upgrading the building to conform to the acceptable performance criteria should be identified and described.

2.4.5 Cost–Impact Study

A cost study, as described in Section 2.3.4, may be required for the detailed assessment.

2.4.6 Recommendations

The results of the detailed assessment shall be summarized in a report (Chapter 5) with a recommendation as to the appropriate course of action. This could include acceptance of the building as is, rehabilitating the building to correct the deficiencies identified, changing the use of the building, or phasing the building out of service.

CHAPTER 3
STRUCTURAL MATERIALS ASSESSMENT

3.1 PURPOSE AND SCOPE

State-of-the-art practices and methods for assessing the condition of selected materials in buildings include visual investigation, nondestructive evaluation (NDE) tests, and destructive tests, including both field and laboratory procedures. The practices and methods for assessing the condition of several building materials are included in Chapter 6 and are described in general terms with reference to documents that provide additional information. The information obtained from using the described practices and methods may assist in the selection of additional tests to be used in the evaluation process. Other appropriate practices and tests should not be excluded from consideration. Chapter 6 includes the assessment of the following commonly used construction materials: concrete, steel, masonry, and wood.

References for the Structural Materials Assessment can be found in Appendix B1 and supplemental references for the Structural Materials Assessment can be found in Appendix B2.

CHAPTER 4

EVALUATION PROCEDURES AND EVALUATION OF STRUCTURAL COMPONENTS AND SYSTEMS

4.1 EVALUATION PROCEDURES

4.1.1 General

Evaluation, as used herein, is defined as the process of determining the structural adequacy of the building or component for its intended use. Evaluation by its nature implies the use of personal judgment by qualified design professionals. Because these design professionals must ultimately take professional responsibility for the evaluation, the evaluation process cannot be standardized. However, guidelines for the evaluation process are established in this section for the purpose of convenience.

4.1.2 Evaluation and Acceptance Criteria

Where practical, the design professional and client should agree on the scope of services necessary for any evaluation. Although most decisions will be based on experience, there may be resources available to guide the design professional.

4.1.3 Recognized Methods

Recognized reliable methods should be used for all evaluation procedures.

4.1.4 New Methods

The use of new or improved evaluation methods is encouraged. When new methods are utilized in evaluation procedures, there should be reliable and sufficient checks of these new methods by the use of recognized methods.

4.2 EVALUATION OF COMPONENTS AND SYSTEMS

4.2.1 Basis of Evaluation

The evaluation should be consistent with the stated purpose(s) of the assessment (Chapter 1, Section 1.2).

4.2.2 Considerations

4.2.2.1 Interpretation of Data. Data should be interpreted statistically when appropriate and feasible.

4.2.2.2 Material Properties. Physical, chemical, and other properties of materials and components may be determined by laboratory or in situ testing.

4.2.2.3 Sample Size Significance. Due consideration and weight should be given to the importance of an element or unit being tested, evaluated, and so on, with respect to the other elements or units in the overall system being evaluated. It is both permissible and desirable to weigh this significance mathematically. A minimum number of samples must be taken to ensure statistical significance and to establish a confidence level, unless taking such sampling will weaken the structure, member, or create unstable conditions. Because a portion of the phi reduction factor is based on variable material properties, a significant number of material tests will eliminate this variability and justify revising the phi reduction factor. Several ASTM standards exist to guide an investigator in determining sample size if a statistically significant sample is desired. Often, a statistically significant sample is not available or economically viable. In this case, it is desired to select a sample without bias. This sample could also be a random sample, but not necessarily, as purely random samples in a failure investigation are often not practical given the constraints of the existing structure. As sample size and distribution are determined, care should be exercised to obtain samples from damaged and undamaged and questionable and unquestionable portions of the structure if applicable.

ASTM E122, *Standard Practice for Calculating Sample Size to Estimate, with a Specified Tolerable Error, the Average for a Characteristic of a Lot or Process* (ASTM 2022) can be followed to determine the number of samples required for a desired level of accuracy. The number of samples will depend on

- Error (E) that can be tolerated in the estimate of the measured value;
- Preferred probability, or confidence level, that the measured value is within the limit of error; and
- Estimate of the variation (σ) of the measured value from one unit to another within the population.

Assuming that the measured result is normally distributed and that the population is large, the following equation, based on an assumed 95%confidence interval, can be used to calculate the number of samples (n) for testing

$$n = \left(\frac{1.96\sigma}{E} \right)^2 \tag{4-1}$$

where

n = Number of samples,
σ = Standard deviation of measured results, and
E = Allowable error in the measured result.

The factor 1.96, based on a desired 95% confidence level, gives a 0.050 probability that the actual value is not within plus or minus the allowable error E of the measured property.

Deterministic sampling describes a situation when a set number of samples are available, or sampling is constrained by any of the factors discussed herein, and only a set number of samples can be taken (e.g., access, effects on the element under consideration, budget). This sampling protocol requires engineering judgment to ensure that a representative estimate of the characteristic under consideration is obtained.

4.2.2.4 Measure of Safety. It is permissible and desirable to use limit state concepts for the determination of a reliability (or safety) index in lieu of a conventional factor of safety. An increase or decrease in the normal factor of safety may be considered. The evaluation should determine an appropriate factor of safety on the basis of the existing loads, the usage, and the type of analysis and investigation conducted.

4.2.2.5 System Approach. Approaches that give cognizance to system evaluation rather than to component or subsystem evaluation should be utilized. The use of load redistribution is acceptable as long as a physical redistribution system is available.

4.2.2.6 Load Tests. A load test provides only a limited amount of information on the performance under a specific load. Many other factors exist that must be considered and evaluated to determine the acceptability of a structure. Load tests should be conducted in conjunction with analytical modeling of the test area and respective verification of the test results. The performance and interpretation of load tests require special knowledge.

Load testing is a procedure that may be used to physically verify the load-carrying capacity of a structural assembly. Such testing procedures must be designed and coordinated and may involve shoring, deflection monitoring, and a mechanism to apply incremental loading to a level in excess of the load capacity to be verified. Referenced procedures are defined in the *International Building Code* (IBC Chapter 17) (ICC 2018), as well as by ACI 318 (ACI 2019). Note that high reliability of the load capacity determined by a load testing procedure applies only to the actual assemblies that have been tested; variations in construction quality or material properties can create a significant variability between the load capacities of areas of similar construction.

4.2.2.7 Stiffness. Stiffness is the resistance to deformation of a member or structure measured by the ratio of the applied force to the corresponding displacement. A change in stiffness of an existing building or a change in the distribution of the stiffening elements, as well as in the stiffness of these individual elements, can materially affect both the distribution of the applied load and the resulting deformations. Overall structural stiffness is made up of many components, including the structural frame, floor systems, exterior walls and panels, architectural in-fill walls, stairs, and composite action of structural and nonstructural elements. Damage or alteration to the components of stiffness, which could be construed to be only architectural in nature, might, in fact, seriously affect the load-carrying capacity of the structure or its component elements.

4.2.2.8 Stability. Stability is the ability of a structure to resist load, while remaining in equilibrium. Loss of stability is indicated by an exhaustion of this resistance. The factor-of-safety against collapse is the ratio of the maximum load a structure can support to the estimated loads. Stability is closely related to stiffness in that many of the components of building stiffness contribute to the overall building stability or factor-of-safety against collapse. Accordingly, those same structural elements that affect stiffness will also affect stability and should be evaluated as part of a structural condition assessment to determine the effects on the building's stability.

4.2.2.9 Overturning and Sliding. Overturning and sliding of structures under the action of lateral loads shall be computed and evaluated against appropriate factors of safety.

4.3 INTERPRETATION

Interpretation of the evaluation results is based on the experience and judgmental opinion(s) of design professionals who are responsible for the evaluation. Because these design professionals will be accepting the professional responsibility for the overall evaluation, interpretation criteria are not standardized herein.

References for this chapter can be found in Appendix B3 and supplemental references for Evaluation Procedures and Evaluation of Components and Systems can be found in Appendix B4.

5

REPORT ON STRUCTURAL CONDITION ASSESSMENT

The report content should be consistent with the agreed-on scope of services. A simple checklist may suffice for a cursory assessment and a letter report for a preliminary assessment. A full report is more appropriate for a detailed assessment.

This chapter is a guide because the order and content of the report may vary with the scope of the engagement, the methods and techniques employed by the design professional, and for specific investigations. It provides a description of the various parts of the preliminary and detailed assessments as may be used. The section as written presents a form for a typical client/consultant relationship, but it does not preclude a different format. Appendix A is an example outline for such a report.

In the case of public agencies or multibuilding owners where routine or special condition assessments are performed by staff personnel for the management of facilities, the form of the report may be established by that agency or owner. Similarly, such a client may require a specific format when the condition assessment is performed by a consultant.

It must be recognized that any report could be used in future litigation. Hence, the degree of investigation and analysis should be stated explicitly. Conclusions and recommendations should be carefully worded, and the possible consequences of not following recommendations should be stated. A carefully worded disclaimer regarding opinions only being based on observations and reviewed materials at the time of preparation of the report may be included. The disclaimer should include a statement that indicates that if additional evidence is discovered, the opinions may change.

5.1 EXECUTIVE SUMMARY

An executive summary at the beginning of the report is discretionary. If included, it should contain brief statements of the purpose, scope, conclusions, and recommendations.

5.2 INTRODUCTION

5.2.1 Purpose of Assessment

The introductory section should be a concise statement describing the reasons for the condition assessment. Background information, if pertinent, may be included in this part and would incorporate applicable government/owner/user reporting requirements (Chapter 1, Section 1.2).

5.2.2 Scope of Investigation

The scope of the investigative work performed for the assessment will vary with the assignment and must be indicated specifically and clearly. Load assumptions and code jurisdictions should be indicated in this portion. Any unusual design features in life safety areas should be given special consideration.

5.2.2.1 Cursory Assessment. A cursory assessment may be used for multiple buildings to initially assess the general condition and for screening to establish priorities (Chapter 2, Section 2.1).

5.2.2.2 Preliminary Assessment. An initial "walk-through" visit for orientation and general impressions is a common part of most preliminary assessments. A review of available documents, further site investigations, preliminary analysis, and preliminary evaluation and recommendations may be part of this work (Section 2.3). Results may indicate the need for an increase in the scope of the investigation.

5.2.2.3 Detailed Assessment. This is an expansion of the preliminary assessment. It includes a further review of documentation, building investigations, materials assessment, detailed analysis, cost impact study, detailed evaluation, and recommendations (Chapter 2, Section 2.4).

5.2.2.4 Survey. The field investigation and observations should be described, including a comparison of as-built to as-designed conditions and field evaluation of existing conditions.

5.2.2.5 Testing. The range and types of testing employed are outlined in this section.

5.2.3 Methods and Techniques

This section includes detailed descriptions of methods and techniques employed in the survey, investigation, testing, visual observations, photographs, oral and video tapes, drawings and sketches, measurements, and investigative procedures and tests.

5.2.4 Meetings

A summary of meetings during the investigative phase should be included.

5.3 DESCRIPTION OF THE STRUCTURE

5.3.1 General

A general description of the structure assessed should be given at this point. This description should include the type of architecture, type of structure, number of stories, orientation of the building on the lot, additions and renovations, occupancy or use classification, and materials comprising the structure.

5.3.2 History

The prior uses or occupancy of the building throughout its life, maintenance procedures, environmental and site conditions, and any unusual loading or other factors should be discussed. In addition, dates of construction, conversion, alteration, and repair should be provided in this section.

5.3.3 Collected Data

Information accumulated during the survey, investigation, and assessment should be listed. This may include original drawings, insurance data, alterations, photographs and tapes, measured drawings, interviews, design calculations, and so on.

5.4 DISCUSSION OF SITE VISIT

5.4.1 Overview

This is a report of the initial site visit for orientation and general impressions. On a limited assignment, this may comprise the entire engagement.

5.4.2 Survey

A more thorough survey should consider materials, real or inferred systems, dimensions, deflections and distortions, identification of problem areas, sample locations, and a record of data obtained.

5.4.3 Observations and Their Significance

This should be a summary of observations made and how they affect the assessment.

5.5 RESEARCH AND ANALYSIS

5.5.1 Computational Analysis

Describe the methods used to analyze the collected information for the assessment of the building structure.

5.5.2 Code Conformance

Compliance with building and fire codes, life safety requirements, and special owner/user criteria should be discussed for the initial construction, alterations, present condition, and potential future use (if applicable).

5.6 TEST PROGRAM

Identify test areas, materials tested, and the appropriate tests performed. Describe in detail the testing methods employed. Results should be included in this section. Nondestructive, destructive, and load tests may be used.

5.7 FINAL COMPUTATIONAL ANALYSIS

This will be part of the basis for the evaluation of the structure.

5.8 INPUT FROM OTHER DISCIPLINES

The results of any surveys or evaluations by other disciplines as they affect the building structure should be given here. Final reports by other disciplines can be attached as appendixes.

5.9 SUMMARY OF THE STUDY

Field and office work directed toward the condition assessment should be summarized.

5.10 CONCLUSIONS

Conclusions are based on the survey, investigation, testing, and evaluation. These require experience and "professional judgment."

5.11 RECOMMENDATIONS

As such, the recommendations are not considered to be part of the guideline, although they are the most important part of the report. Prepare a recommended repair program if appropriate for maintaining the structural adequacy of the building.

5.12 APPENDIXES

These should include all supporting data such as survey information, record drawings, photographs, test data and reports, computations, and references.

CHAPTER 6
COMMON BUILDING MATERIALS

6.1 CONDITION ASSESSMENT OF CONCRETE

The condition of a concrete structure is best determined through an understanding of its material properties and physical conditions by visual examination in conjunction with applicable computational methods and testing. Major material properties and physical conditions are discussed in Section 6.1.4. Test methods that can be used to evaluate these properties and physical conditions are provided in Section 6.1.13. A condition assessment of concrete may require determinations of strength and quality. Proper assessment and subsequent evaluation (Section 6.1.1) should be utilized to reveal the concrete's ability to sustain loads and endure the impacts of the environmental conditions to which it is being or will be subjected.

Basic information for conducting an engineering analysis of existing structural concrete requires knowledge about how new structures are designed. The code-adopted edition of ACI 318, *Building Code Requirements for Reinforced Concrete and Commentary* (ACI 2019a), should be referenced for this purpose. Older, existing structures may require reference to earlier editions of ACI 318 for previously used design methods and requirements. Information for conducting an engineering evaluation can be found in ACI 562, *Code Requirements for Assessment, Repair, and Rehabilitation of Existing Concrete Structures and Commentary* (ACI 2021); ACI 201.1R, *Guide for Conducting a Visual Inspection of Concrete in Service* (ACI 2008); ACI 228.2, *Report on Nondestructive Test Methods for Evaluation of Concrete in Structures* (ACI 2013); ACI 364.1R, *Guide for Assessment of Concrete Structures before Rehabilitation* (ACI 2019b); ACI 437.2, *Code Requirements for Load Testing of*

Existing Concrete Structures and Commentary (ACI 2023); and ACI 563, *Specifications for Repair of Concrete in Buildings* (ACI 2018).

6.1.1 Reasons for Conducting a Condition Assessment of Concrete

A structural condition assessment may be conducted on a concrete structure for a number of reasons. The most common *technical* motivations for an assessment often relate to one or more of the following categories:

Changes in Usage, Occupation, or Codes: These can result in the need for structural modifications or other improvements on account of different loading conditions, additional safety requirements, and so on.

Moisture Levels: Determining moisture levels is necessary to assess the impact of moisture content and to determine the sources of the moisture to better understand the concrete's durability, reasons for deterioration, and to gauge future performance.

Observed Deterioration: Comprehensive survey of observable deterioration is needed to determine the cause, locate and quantify the extent, and assess its impact on material properties.

Strength or Stiffness Questions: There is a need to determine the strength and stiffness of individual members or components to understand their capacity and mechanical properties and evaluate their adequacy for existing or proposed loads.

Structural Capacity and Performance: There is a need to determine or verify by calculation or other means the ability of the concrete member or structure to carry load and perform as intended and is required.

Environmental Damage: There is a need to understand the effects on concrete from observed loads caused by wind, earthquakes, flooding, and so on.

Note: Making modifications to existing buildings to increase their energy efficiency, such as adding insulation in exterior walls or at roofs, can alter the moisture and temperature environment of a building. Any modifications or alterations to the original building may warrant additional consideration when conducting a concrete condition assessment.

6.1.2 Concrete as a Material

Concrete is a popular building material composed of a mixture of cementitious material, water, aggregates, and admixtures. When the concrete is mixed, it is in a liquid, or plastic, state. The cement then undergoes a hydration reaction that causes the mixture to harden and develop strength. Concrete is placed in its liquid state, making it free to fill the extent of the formwork (freedom of geometry). The mixture is initially mixed and placed in a liquid or plastic condition, which when allowed to set transitions to a solid state.

Plain concrete (without reinforcement) has low tensile and shear capacities relative to its compressive strength. This makes it especially vulnerable to lateral loads caused by wind, earthquakes, vibrations, and other forces and is therefore unsuitable in many structural applications. The nineteenth-century invention of reinforced concrete, which allows concrete's tensile and shear capacities to be supplemented by reinforcing materials with higher tensile and shear capacities, has revolutionized the use of concrete and has allowed it to become one of the world's most common building materials.

In reinforced concrete, the reinforcement, usually steel, is embedded in the concrete. The two materials are then able to work compositely to allow the member to sustain imposed stresses, with concrete providing the majority of the compressive strength and the reinforcing typically providing the majority of the shear and tensile strength. Reinforcing, such as rods, bars, fibers, or mesh, contributes in resisting the tensile, shear, bending, and compressive stresses. Unreinforced and reinforced concrete members and structures include cast-in-place and precast concrete, prestressed concrete (pretensioned or posttensioned), or combinations thereof.

The function of concrete as a material in a structure is twofold. First, the concrete functions as one component of the composite structural member that constitutes the load-carrying element. Second, the concrete provides protection against fire and environmental forces. Using a denser, higher-quality concrete, providing adequate cover, and minimizing cracking can delay and reduce corrosion of the embedded steel reinforcing as well as act as a barrier to reduce the ingress of deleterious agents.

The function of reinforcing steel, pretensioning steel, posttensioning steel, and cable in a concrete structure is to carry and transmit either compressive or tensile stresses. The properties and physical conditions of the steel need to be determined to evaluate the stress-carrying ability of the concrete structure (i.e., the interface between the concrete and the embedded reinforcement).

For concrete members or elements to function as a load-carrying structural element, the following two coincidental characteristics are required: adequate compressive strength and adequate cross-sectional size (which contributes to member stiffness). If either or both of these characteristics is not adequate, the member or element may require remediation.

6.1.3 Terminology Related to Concrete Characteristics and Properties

Whether a concrete product is cast-in-place or precast, reinforced or plain, the basic terminology relevant to condition assessments is the same.

For the sake of brevity, rather than reproduce in its entirety a list of all relevant terminology within this chapter, an abbreviated list of common terms is included in the following. The ACI Manual of Practice, ACI Concrete Terminology, and ACI 116R provide extensive lists of cement and concrete terminology and are excellent sources for this and more information. A few definitions that can be of assistance in the assessment of buildings are included as follows:

Blistering: Irregular raising of a thin layer at the surface of placed mortar or concrete during or soon after completion of the finishing operation.

Breakout Strength, Concrete: Strength of the concrete required to breakout a volume of concrete surrounding the anchor or group of anchors under a load on the anchor or group of anchors.

Bug Holes: Small regular or irregular cavities, usually not exceeding 15 mm (0.6 in.) in diameter, resulting from entrapment of air bubbles at the surface of formed concrete during placement and consolidation (also known as *surface air voids*).

Cementitious Materials: Materials that have cementing value if used in concrete either by themselves, such as portland cement, blended hydraulic cements, and expansive cement, or such materials in combination with fly ash, other raw or calcined natural pozzolans, silica fume, and slag cement.

Chalking: Formation of a loose powder resulting from the disintegration of the surface of concrete or an applied coating, such as cementitious coating.

Cold Joint: Joint or discontinuity resulting from a delay in placement of sufficient duration to preclude intermingling and bonding of the material in two successive lifts of concrete, mortar, or the like.

Concrete: Mixture of portland cement or any other cementitious material, fine aggregate, coarse aggregate, and water, with or without admixtures.

Concrete Strength, Specified Compressive: Compressive strength of concrete used in the design and evaluated in accordance with provisions of this code, psi; wherever the quantity f_c' is under a radical sign, the square root of the numerical value is only intended, and the result has units of psi.

Concrete, NonPrestressed: Reinforced concrete with at least the minimum amount of nonprestressed reinforcement or no prestressed reinforcement, and for two-way slabs, those with less than the minimum amount of prestressed reinforcement.

Concrete, Plain: Structural concrete with no reinforcement or with less than the minimum amount of reinforcement specified for reinforced concrete.

Concrete, Precast: Structural concrete element cast elsewhere than its final position in the structure.

Concrete, Prestressed: Reinforced concrete in which internal stresses have been introduced by prestressed reinforcement to reduce potential tensile stresses in concrete resulting from loads, and for two-way slabs, with at least the minimum amount of prestressed reinforcement.

Concrete, Reinforced: Structural concrete reinforced with at least the minimum amount of nonprestressed reinforcement, prestressed reinforcement, or both, as specified in this guide.

Concrete, Steel Fiber-Reinforced: Concrete containing a prescribed amount of dispersed, randomly oriented, discontinuous deformed steel fibers.

Corrosion: Deterioration of a material, usually metal, that results from a chemical reaction with its environment.

Cover, Specified Concrete: Distance between the outermost surface of embedded reinforcement and the closest outer surface of the concrete.

Crack: A complete or incomplete separation, of either concrete or masonry, into two or more parts produced by breaking or fracturing.

Craze Cracks: Fine random cracks or fissures in a surface of plaster, cement paste, mortar, or concrete.

Crazing: The development of craze cracks; the pattern of craze cracks existing in a surface.

Curling: The distortion of a concrete member from its original shape, such as the warping of a slab caused by differences in temperature or moisture content in the zones adjacent to its opposite faces.

D-cracks: Series of cracks in concrete near and roughly parallel to joints and edges.

Deflection: Movement of a point on a structure or structural element, usually measured as a linear displacement or as succession displacements traverse to a reference line or axis.

Deformation: Change in dimension or shape.

Delamination: Separation along a plane parallel to a surface, as in the case of a concrete slab, a horizontal splitting, cracking, or separation within a slab in a plane roughly parallel to, and, in general, near, the upper surface; found most frequently in bridge decks and caused by the corrosion of reinforcing steel or freezing and thawing; similar to spalling, scaling, or peeling except that delamination affects large areas and can often be detected only by nondestructive tests such as tapping or chain dragging.

Deterioration: (1) physical manifestation of failure of a material (e.g., cracking, delamination, flaking, pitting, scaling, spalling, and staining) caused by environmental or internal autogenous influences on rock and hardened concrete as well as other materials, or (2) decomposition of material during either testing or exposure to service.

Diagonal Crack: Diagonal crack through a flexural member, an inclined crack, caused by shear stress, usually at approximately 45 degrees to the axis, or a crack in a slab, not parallel to either the lateral or the longitudinal directions.

Discoloration: Departure of color from that which is normal or desired.

Disintegration: Reduction into small fragments and subsequently into particles.

Drop Panel: Projection below the slab used to reduce the amount of negative reinforcement over a column or the minimum required slab thickness and to increase the slab shear strength.

Duct: Conduit, plain or corrugated, to accommodate prestressing reinforcement for posttensioning applications.

Durability: Ability of a structure or member to resist deterioration and affect the structure's useful life.

Dusting: Development of a powdered material at the surface of hardened concrete.

Edge Distance: Distance from the edge of the concrete surface to the center of the nearest anchor.

Efflorescence: Deposit of salts, usually white, formed on a surface, the substance having emerged in solution from within either concrete or masonry and subsequently been precipitated by a reaction, such as carbonation or evaporation.

Embedments: Items embedded in concrete. Reinforcements or anchors welded, bolted, or otherwise connected to the embedded item to develop the strength of the assembly are considered to be part of the embedment.

Exfoliation: Disintegration occurring by peeling off in successive layers, swelling up, and opening into leaves or plates like a partly opened book.

Hairline Cracks: Cracks in an exposed-to-view concrete surface having widths so small as to be barely perceptible.

Honeycomb: Voids left in concrete because of failure of the mortar to effectively fill the spaces among coarse aggregate particles.

Incrustation: A crust or coating, typically hard, formed on the surface of concrete or masonry construction or aggregate particles.

Joint Fault: Differential displacement of a portion of a structure along a joint.

Laitance: Layer of weak material known as *residue* derived from cementitious material and aggregate fines either (1) carried by bleeding to the surface or the internal cavities of freshly placed concrete or (2) separated from the concrete and deposited on the concrete surface or internal cavities during placement of concrete underwater.

Spall: Perhaps roughly a circular depression not greater than 20 mm (0.8 in.) in depth and 150 mm (6 in.) in any dimension.

Leakage: Contained material migrating through the concrete member.

Longitudinal Cracks: Crack that develops parallel to the length of the member.

Map Cracking: (1) Intersecting cracks that extend below the surface of hardened concrete; caused by shrinkage of the drying surface concrete that is restrained by concrete at greater depths where either little or no shrinkage occurs; vary in width from fine and barely visible to open and well-defined, or (2) the chief symptom of a chemical reaction between alkalis in cement and mineral constituents in aggregate within hardened concrete, caused by the differential rate of volume change in different members of the concrete; cracking is usually random and on a fairly large scale and, in severe instances, the cracks may reach a width of 12.7 mm (0.50 in.).

Modulus of Elasticity: Ratio of normal stress to corresponding strain for tensile or comprehensive stresses below the proportional limit of material.

Pattern Cracking: Cracking on concrete surfaces in the form of a repeated sequence, resulting from a decrease in volume of the material near the surface, or an increase in volume of the material below the surface, or both.

Peeling: Process in which thin flakes of mortar are broken away from a concrete surface, such as by deterioration or by adherence of surface mortar to forms as forms are removed.

Pitting: Development of relatively small cavities in a surface; in concrete, localized disintegration, such as a popout; in reinforcing or prestressing steel, localized corrosion evident as minute cavities on the surface.

Plastic Shrinkage Cracking: Cracking that occurs in the surface of fresh concrete soon after it is placed and while it is still plastic.

Popout: Breaking away of small portions of a concrete surface because of localized internal pressure that leaves a shallow, typical conical depression with a broken coarse aggregate at the bottom.

1. *Popouts, Large*: Popouts leaving depressions greater than 50 mm (2 in.) in diameter.
2. *Popouts, Medium*: Popouts leaving depressions between 10 and 50 mm (0.4 and 2 in.) in diameter.

3. *Popouts, Small*: Popouts leaving depressions up to 10 mm (0.4 in.) in diameter, or the equivalent.

Posttensioning: Method of prestressing in which prestressing reinforcement is prestressed after concrete is cast and cured.

Pretensioning: Method of prestressing in which prestressing reinforcement is tensioned before concrete is cast.

Reinforcement, Deformed: Deformed bars, welded bar mats, deformed wire, and welded wire reinforcement, excluding plain wire.

Reinforcement, Nonprestressed: Bonded reinforcement that is not prestressed.

Reinforcement, Plain: Bars or wires that do not conform to the definition of deformed reinforcement.

Reinforcement, Prestressed: High-strength reinforcement, such as strand, wire, or bar that has been tensioned to impart forces on concrete.

Reinforcement, Welded Deformed Steel Bar Mat: Mat consisting of two layers of deformed bars at right angles to each other welded at the intersections.

Reinforcement, Welded Wire: Plain or deformed wire fabricated into sheets or rolls.

Sand Pocket: Zone in concrete or mortar containing fine aggregate with little or no cement material.

Sand Streak: Streak of exposed fine aggregate in the surface of formed concrete caused by bleeding.

Scaling: Local flaking or peeling away of the cement matrix near-surface portion of hardened concrete or mortar.

1. *Scaling, Light*: Loss of surface mortar without exposure, of course, aggregate
2. *Scaling, Medium*: Loss of surface mortar 5 to 10 mm (0.2 to 0.4 in.) in depth and exposure of coarse aggregate.
3. *Scaling, Severe*: Loss of surface mortar 5 to 10 mm (0.2 to 0.6 in.) in depth with some loss of mortar surrounding aggregate.
4. *Scaling, Very Severe*: Loss of coarse aggregate particles as well as surface mortar, typically to a depth greater than 20 mm (0.8 in.).

Segregation: Differential concentration of the components of mixed concrete, aggregate, or the like, resulting in nonuniform proportions in the mass.

Sheathing: Material encasing prestressing reinforcement to prevent bonding of the prestressing reinforcement with the surrounding concrete, to provide corrosion protection, and to contain the corrosion-inhibiting coating.

Shrinkage Cracking: Cracking of a structure or member on account of failure in tension caused by external or internal restraints as a reduction in moisture content develops, carbonation occurs, or both.

Spall: Fragment, usually in the shape of a flake, detached from a concrete member by a blow, by the action of weather, by pressure, by fire, or by expansion within the larger mass.

Spall, Small: A roughly circular depression not greater than 20 mm (0.8 in.) in depth and 150 mm (6 in.) in any dimension.

Spiral Reinforcement: Continuously wound reinforcement in the form of a cylindrical helix.

Staining: Discoloration by foreign matter.

Stalactite: Downward-pointing deposit formed as an accretion of mineral matter produced by the evaporation of dripping liquid from the surface of concrete, commonly shaped like an icicle.

Stalagmite: Upward-pointing deposit formed as an accretion of mineral matter produced by the evaporation of dripping liquid, projecting from the surface of rock or of concrete, commonly roughly conical in shape.

Stirrup: Reinforcement used to resist shear and torsion forces in a member; typically, deformed bars, deformed wires, or welded wire reinforcement either single leg or bent into L, U, or rectangular shapes and loaded perpendicular to, or at an angle to, longitudinal reinforcement.

Stratification: Separation of overwet or overvibrated concrete into horizontal layers with increasingly lighter material toward the top; water, laitance, mortar, and coarse aggregate tend to occupy successively lower positions in that order; a layered structure in concrete resulting from the placing of successive batches that differ in appearance; occurrence in aggregate stockpiles of layers of differing grading or composition; a layered structure in a rock formation.

Structural Performance: Structure's integrity and safety may be evident because of excessive cracking or other indications of structural distress. Poor or inadequate performance can result from design deficiencies, construction-related causes, excessive applied loads, or deformations caused by damage from fire or other external or environmental means.

Temperature Cracking: Cracking caused by tensile failure, caused by temperature drop in members subjected to external restraints, or caused by a temperature differential in members subjected to internal restraints.

Tendon: In posttensioned members, a tendon is a complete assembly consisting of anchorages, prestressing reinforcement, and sheathing with coating for unbonded applications or ducts filled with grout for bonded applications.

Tendon, Bonded: Tendon in which prestressed reinforcement is continuously bonded to the concrete through grouting of ducts embedded within the concrete cross section.

Tendon, External: Tendon external to the member concrete cross section in posttensioned applications.

Tendon, Unbonded: Tendon in which prestressed reinforcement is prevented from bonding to the concrete. The prestressing force is permanently transferred to the concrete at the tendon ends by the anchorages only.

Tie: (a) Loop of reinforcing bar or wire enclosing longitudinal reinforcement; a continuously wound transverse bar or wire in the form of a circle, rectangle, or other polygonal shapes without reentrant corners enclosing longitudinal reinforcement; and (b) a tension element in a strut-and-tie model.

Transverse Cracks: Cracks that occur across the longer dimension of a member.

Uniformity of Concrete: Degree or consistency of the properties of the concrete from one part of the structure to another.

Warping: Out-of-plane deformation of the corners, edges, and surface of pavement, slab, or wall panel from its original shape.

Water–Cementitious Materials Ratio: Ratio of mass of water, excluding that absorbed by the aggregate, to the mass of cementitious materials in a mixture, stated as a decimal.

Yield Strength: Specified minimum yield strength or yield point of reinforcement; yield strength or yield point shall be determined in tension according to applicable ASTM standards.

6.1.4 Physical and Mechanical Properties of Concrete

As a structural material, the properties and physical conditions of concrete should be considered in evaluating the acceptability of existing concrete. The determination of the concrete properties may be required to provide an indication of its condition.

As structural materials, the properties and physical conditions of reinforcing steel, pretensioning steel, posttensioning steel, and cables that are summarized in Tables 6-5 and 6-6 should be considered in evaluating the acceptability of existing reinforcement.

Observed problems in a structure often serve as good indicators of areas requiring investigation. Commonly measured properties that can be used to investigate and analyze the condition of a concrete member or structure are defined and discussed in this section.

Absorption: Represents the increase in mass of concrete resulting from the penetration of water into the pores. It is usually measured after submerging a concrete specimen in water for a period of time. This property is considered to be a good predictor of the durability of concrete.

Acidity: Concrete is chemically basic, normally having a pH of around 12 to 13, which provides a passive environment for the steel reinforcement. When the concrete is carbonated or is exposed to acids (pH values less than 7), the concrete and reinforcement can actively deteriorate.

Air Content: Volume of air voids in concrete is usually expressed as a percentage of total volume of the paste, mortar, or concrete.

- The American Concrete Institute has prepared a list of more than 250 chemicals documenting their effect on concrete (ACI 5, ACI 15.1).
- Concrete that exhibits damage from freezing and thawing will most likely have an air content below the percentage recommended (ACI 318, ACI 350R, and ACI 201.1R).

Carbonation: Occurs when calcium hydroxide (lime) in cement, mortar, or concrete reacts with carbon dioxide in the air to form calcium carbonate and, over time, reduces the pH of the concrete. When the pH reaches 9.5 or lower, the passive protection on steel reinforcing provided by the concrete's alkalinity diminishes, and corrosion can initiate. Figure 6-1 depicts the use of phenolphthalein to identify carbonated areas during a petrographic analysis of concrete samples (purple areas are unaffected by carbonation).

Cement Content: Relative mass of cement in concrete is an important factor that can affect other properties such as the heat generated during hydration and hence the susceptibility of the concrete to shrinkage. The durability of concrete also depends on the impermeability of the cement paste. One way to ensure the necessary degree of impermeability of the paste is to specify a minimum cement content for a mixture (Mather 1989).

Chemical Content: Presence and amount of foreign compounds within the concrete. Most foreign compounds in concrete result in a reduced compressive strength or deterioration. As an example, calcium chloride, which has often been used beneficially as an accelerator, in excess amounts can be detrimental to reinforced concrete by inducing corrosion of reinforcing steel (ACI 318, ACI 222.1R).

Chloride Content: Represents the mass of free chloride in the concrete and is a significant measure of active corrosion of reinforcing steel. Reinforced concrete exposed to chloride ions rapidly deteriorates once the chloride content exceeds the corrosion threshold limit. This threshold is best measured by the ratio of chloride ions to hydroxyl ions. The diffusion of the chloride ions through concrete

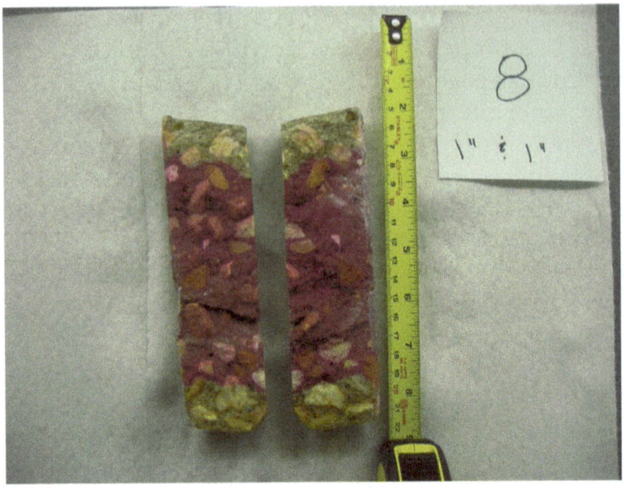

Figure 6-1. Carbonation testing of concrete nonpurple areas of cores is considered carbonated pH < 9.5.
Source: Courtesy of Vector Corrosion Technologies.

follows Fick's law, which is a function of time, the chloride concentration at the surface, and a diffusion constant (Cady and Weyers 1983, Weyers and Smith 1989). Concrete samples can be collected at various depths with roto-hammer drills to determine the depth of chlorides exceeding the threshold limit. Figure 6-2 depicts the collection of concrete dust for chloride content testing.

Compressive Strength: Measured maximum resistance to axial compressive loading expressed as force per unit cross-sectional

Figure 6-2. Chloride content test sampling with a roto-hammer drill.
Source: Courtesy of Vector Corrosion Technologies.

area of the concrete. It is one of its most important properties, not only because compressive strength is how concrete is designated but also because it is, directly and indirectly, related to many other properties. These include density, durability, modulus of elasticity, modulus of rupture, permeability, and soundness. Concrete compressive strength is determined by a standardized testing procedure. It can represent the value that concrete was originally designed for or what it is determined to presently be for an investigation. Any investigation of concrete strength to evaluate its design strength should attempt to correlate investigative data of tests on specimens taken from the hardened concrete to a standardized test. This means that results must be corrected for sample size differences (Section 6.1.13.2, Table 6-2, ASTM C42/C42M), for maturity (Section 6.1.13.2, Table 6-2, ASTM C1074) (Plowman 1956), and statistically (Section 6.1.13.2, Table 6-2, ACI 214.3R).

Creep: Dimensional increase caused by strain under sustained loading. A total of 90% of creep occurs in the first two years of life (ACI 318, ACI 209). It can typically be assumed that most, if not all, creep deflection has occurred in older concrete structures. However, additional creep can occur if the sustained loads applied to the structure increase.

Elongation: Represents the increase in length of a specimen as a percentage of the original length.

Modulus of Elasticity: Ratio of normal stress to corresponding strain for tensile or compressive stresses below the proportional limit of the material. Because of the relationship between concrete density and concrete compressive strength with the modulus of elasticity, the determination of density and compressive strength is usually sufficient to approximate the modulus of elasticity.

Modulus of Rupture: Determination of the flexural strength of plain concrete; used as a measure of the stress at which tension cracks under flexure will begin to develop (*see* Tensile Strength).

Moisture Content: Percent by mass of free water, an important factor affecting concrete volume changes and often the strength. Concrete volume changes in long-span structures may be significant and may have to be considered in a structural investigation.

Permeability: Measurement of the capacity for movement of water through concrete.

Proportions of Aggregate: Respective proportions of the fine and coarse aggregate in concrete by total concrete mass.

Pullout Strength: Strength determined by measuring the maximum force required to pull an inserted material or element from the hardened concrete.

Resistance to Freezing and Thawing: Used to determine the effect of concrete properties on the resistance of concrete to freezing-and-thawing cycles; may be used as a measure of concrete durability or deterioration.

Soundness: Freedom of a solid, such as concrete, from cracks, flaws, fissures, or variations from an accepted standard.

Splitting Tensile Strength: Indirect method of applying tension to a specimen to cause failure by splitting; it is used as a measure of both diagonal tension capacity and tensile strength.

Tensile Strength: Maximum unit stress that a material is capable of resisting under axial tensile loading, based on the cross-sectional area of the specimen prior to loading. Because of the difficulty of testing concrete in tension, the splitting tensile strength and the modulus of rupture are both used as indirect measures of tensile strength. However, direct procedures are available (Hughes and Chapman 1966, Evans and Marathe 1968, Gopalaratnam and Shah 1985).

Uniformity of Mix: Determines the homogeneity of a concrete mix, to include the cement type, water content, aggregate size and proportions, and the dispersal of any other ingredients throughout the concrete.

Unit Weight: Weight per unit of concrete volume. Normal weight concrete has an air dry unit weight equal to or exceeding 145 lb/ft^3 (2,323 kg/m^3). Lightweight concrete has an air–dry unit weight not exceeding 115 lb/ft^3 (1,850 kg/m^3).

Water–Cement Ratio: Ratio by mass of mixing water to cementitious materials, a significant indicator of concrete strength and durability.

6.1.5 Concrete Products

Concrete can either be placed in situ into a site-specific formwork and allowed to harden in place (cast-in-place) or be fabricated and erected into its final place in the completed structure (precast). Cast-in-place concrete structures are homogeneous, whereas precast concrete structures consist of individual building elements that are attached to each other, typically with steel embed connections. Precast members can be fabricated at the site, for example, tilt-up construction, or at a precast plant. The following are types of concrete members and structures.

Concrete products consisting of structural members and elements can be cast-in-place or precast. Reinforcement products can be made of steel, carbon fiber, glass fiber, or other. Miscellaneous products consist of attachments, anchors, and others.

Cast-in-Place Concrete

Conventionally Reinforced: Concrete poured into a site-specific formwork with conventional reinforcement such as rebar and wire mesh.

Fiber Reinforced: Concrete poured into a site-specific formwork with fiber reinforcement, which can be glass, synthetic, natural, or steel.

Posttensioned Reinforced: Concrete poured into a site-specific formwork with sheathed tensioning cables, which are stressed after the concrete has reached a minimum compressive strength. Conventional reinforcement is also present within posttensioned concrete.

Precast Concrete

Conventionally Reinforced: Concrete poured into forms at an off-site plant with conventional reinforcement such as rebar and wire mesh.

Fiber Reinforced: Concrete poured into forms at an off-site plant with fiber reinforcement, which can be glass, synthetic, natural, or steel.

Pretensioned Reinforced: Concrete poured into forms at an off-site plant with pretensioned cables and conventional reinforcement.

Precast concrete members can be either structural or architectural elements. A limited list of precast structural members consists of columns; inverted, ledger, and rectangular beams; double and bulb tees; hollow-core and solid slabs; load-bearing walls; and box girders. A limited list of precast architectural elements consists of spandrel panels, column covers, pavers, and solid or sandwich wall panels.

6.1.6 Mechanisms of Concrete Deterioration

For information about durable concrete, refer to the code-adopted version of ACI and other relevant ACI documents.

Alkali–Carbonate Reaction: Reaction between alkalis (sodium and potassium) in portland cement and certain carboniferous rocks or minerals such as calcific dolomites and dolomitic limestones. Products of the reaction may cause an abnormal expansion and cracking of concrete in service. The reaction may occur internally between the cement and the aggregates in the concrete and/or externally between the cement and the carboniferous rocks or minerals in the surrounding environment, causing deterioration and damage to the concrete.

Problems associated with dolomite limestones are uncommon, and most limestone aggregates used in concrete have good performance records and are innocuous. Reactive dolomite limestones are very fine-grained, dense, and frequently contain illite and chlorite clay minerals (Hobbs 1988, Appendix B3).

Alkali–Silica Reaction (ASR): Reaction between the alkalis (sodium and potassium) in portland cement and certain siliceous rocks or minerals such as opaline chert and acidic volcanic glass present in some aggregates. Products of the reaction may cause an abnormal expansion and irregular spider web or a rectangular patterned cracking of the concrete (Figure 6-3). The reaction usually occurs internally between the cement and the aggregate in the concrete and could result in a loss of cement paste. It can also occur externally between the cement and the siliceous rocks and minerals in the surrounding environment (such an occurrence is uncommon but possible, for instance, in underground environments).

Alkali–silica reaction in new concrete is usually limited and controlled by the use of nonreactive aggregates, appropriate amounts of pozzolans, and use of low-alkali portland cement. Concrete exposed to saltwater or a

Figure 6-3. Alkali–silica reaction distress.
Source: Courtesy of Raths, Raths, & Johnson, Inc.

salt-water atmosphere over a long period can absorb sodium ions, leading to an eventual alkali–silica reaction.

In existing concrete with ASR, the reaction can be slowed by limiting or preventing external moisture or water from permeating the concrete by the use of surface barrier systems or by the introduction of adequate concentrations of lithium ions into the concrete. Lithium can be introduced by various means with varying effectiveness such as simply by surface wetting with a lithium solution. Other methods to introduce lithium ions include vacuum injection or, most promising by electrochemical treatment.

Bleeding Channels: Internal passageways to the surface of the concrete letting excess water escape during drying and settlement; bleeding channels weaken the concrete and allow water, salts, and other chemicals to attack the concrete interior.

It has been suggested that a greater bleeding capacity of the concrete decreases the plastic shrinkage and that shrinkage cracking takes place when the rate of evaporation exceeds the rate at which bleeding water rises to the surface (ACI Monograph No. 6 1971).

Carbonation: Chemical reaction between calcium hydroxide in hardened concrete and atmospheric carbon dioxide, which can induce corrosion of reinforcing steel (Section 6.1.4.4).

Cement–Aggregate Reaction: Reaction between cement and constituents of aggregate, other than alkali–silica and alkali–carbonate reactions, which include hydration of anhydrous sulfates, rehydration of zeolites, wetting of clays, and reactions involving solubility, oxidation, sulfates, and sulfites. The products of these internal reactions may cause an abnormal expansion and cracking of concrete in service.

Research indicates that the cement–aggregate reaction is mainly a reaction between the alkalis in the cement that produce high pH and abundant hydroxyl and siliceous constituents of the aggregates. Concretes made with reactive sand gravels also have the potential to react.

Chemical Degradation: Degradation of concrete on account of chemical attacks such as sulfate attack, acid attack, and alkali–aggregate reactions (Section 6.1.4.2).

Chloride Attack: Chlorides and nitrates of ammonium, magnesium, aluminum, and iron attack concrete, with those of ammonium being the most harmful. For instance, sodium chloride is chemically harmless to concrete but is a major contributor to corrosion of reinforcing steel.

The primary modes of penetration of chloride through concrete are through cracks and by diffusion (Cady and Weyers 1983).

Contaminated Aggregate: Presence of impurities that may either interfere with the chemical reactions of hydration or induce a chemical degradation of concrete.

Aggregate can be contaminated locally by soils or other environmental exposures. Inadequate washing of the aggregate to remove organic materials and environmental deposits commonly results in contaminated aggregates.

Contaminated Mixing Water: Among other possibilities, impurities in the mixing water may interfere with the chemical reactions of hydration, may adversely affect the strength of concrete, cause staining of the surface and concrete damage, and may also lead to corrosion of the reinforcement.

Cracking: Separation of concrete into parts and is characterized by length, direction, width, and depth, and whether the crack is active or passive. Passive cracks may be caused by construction errors, shrinkage, variations in internal temperature, or shock waves. Active cracks may be caused by variations in atmospheric or internal temperature, absorption of moisture, reinforcement corrosion, chemical reactions, settlement, or various loading conditions.

Research evidence strongly suggests that the bond between mortar and aggregate is the weak link in the heterogeneous concrete system. This relative weakness of the interface between cement paste or mortar and aggregates may be the source of microcracking. The onset of cracking arises mainly from the spread of microcracks along the boundaries of the coarse aggregate (Shah and Winter 1968, ACI Monograph No. 6 1971).

Cross-Sectional Properties: Dimensions and other geometric properties of structural concrete members and components. Testing equipment such as ground-penetrating radar, Figure 6-4, is used to define the thickness of slabs, location of reinforcement, but is limited in finding inclusions within the concrete (voids or delaminations).

Delamination: Horizontal splitting, cracking, or separation of a concrete member in a plane roughly parallel to and typically near the surface. Although it can be found in all types of concrete structures, it is frequently found in bridge and parking deck slabs and is likely caused by the corrosion of the reinforcing steel. It is not readily apparent without testing such as soniscoping (Figure 6-5), chain-dragging, or hammer tapping (Figure 6-6). Delamination usually occurs in the plane of the reinforcing steel.

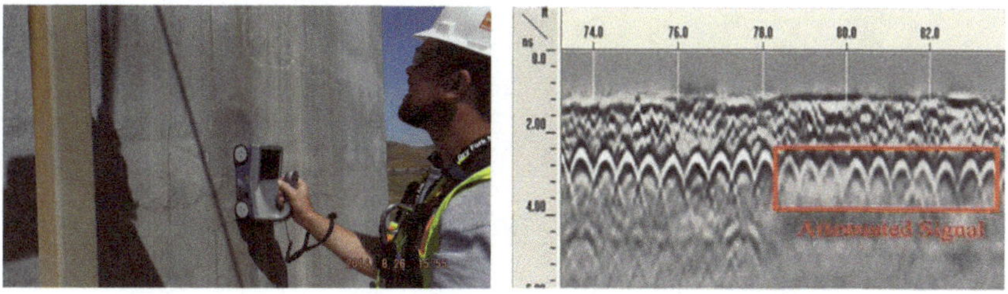

Figure 6-4. Ground-penetrating rebar.
Source: Courtesy of Vector Corrosion Technologies.

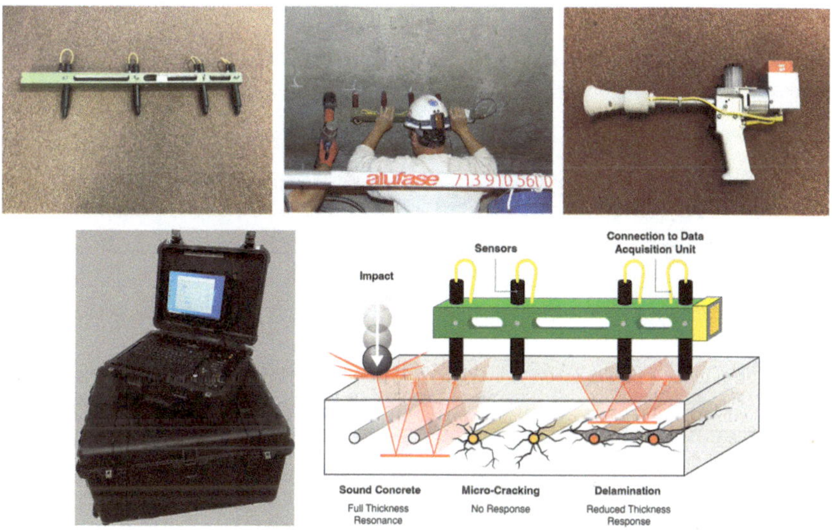

Figure 6-5. Impact ECHO testing to identify voids/delaminations in concrete.
Source: Courtesy of Vector Corrosion Technologies.

Deterioration: Physical impairment of appearance, service life, and the usefulness of the concrete.

Discoloration: Departure of color from that which is normal or desired; usually caused by stains such as from efflorescence and spilled materials or form release agents and chemical changes often resulting from hydration, which is less if calcium chloride-based admixtures are not used.

Disintegration: Deterioration into small fragments or particles caused by factors such as chemical attack, weathering, erosion, and debonding of aggregate and cement paste.

Figure 6-6. Delaminations marked by chalk during a sounding survey. Source: Courtesy of Vector Corrosion Technologies.

Distortion: Warping or deforming of concrete caused by factors such as overloading, poor design, poor construction, ground movement, and expansion.

Efflorescence: Deposit of white salts, usually calcium carbonate, on the concrete surface. For example, Figure 6-7 shows cracking caused by the alkali–silica reaction and resulting efflorescence on the face of the wall.

Erosion: Progressive surface disintegration of concrete by the abrasive or cavitation action of gases, fluids, or solids in motion.

Freeze–Thaw Damage: Surface damage to concrete resulting from the actions of freeze–thaw cycles.

A view of the results of freeze–thaw damage is shown in Figure 6-8.

Concrete may crack because of internal expansion caused by the freezing of trapped water in the capillary cavities in hardened cement paste. This will not happen if the cement paste is provided with a system of entrained air voids so that the bubble spacing factor of paste is 0.008 in. (0.2 mm) or less (Mather 1968).

Also, certain aggregates are susceptible to frost attack (ACI 201.1 and ACI 201.2).

Honeycomb: Voids left in concrete because of a lack of adequate vibration, very low slump, or congestion of the reinforcing steel.

Leaching: Transport of chemicals via water migrating through the concrete. The water dissolves calcium hydroxide or alkalis within the concrete matrix that react with carbon dioxide in the environment, forming calcium carbonate or efflorescence (Section 6.1.6.17) on the concrete surface. Usually this is most evident along cracks or at joints.

Figure 6-7. Efflorescence from an alkaline silica reaction (arrows).
Source: Courtesy of Raths, Raths, & Johnson, Inc.

Popout: Breaking away of small portions of a concrete surface because of internal pressure, which leaves a shallow, typically conical depression. Popouts may be caused by reinforcing steel corrosion, cement–aggregate reactions, or internal ice crystal formation.

Popouts range in diameter and depth. Popouts are measured by the number per ft^2 (m^2) (SHRP-P-338, 1993).

A typical popout is shown in Figure 6-9.

Scaling: Local flaking or peeling away of the cement matrix near-surface portion of hardened concrete or mortar. Scaling is a disintegration process.

Scaling is shown in Figure 6-10.

Spalling: Detachment of fragments (spalls), often resulting from separation of near-surface delaminations from a concrete mass. Variations in internal temperature, corrosion of reinforcement bars, chemical reactions, weathering, poor design, and for other reasons can cause active spalling. Shock waves or a single incident of varying internal temperature can cause passive spalling. Passive

Figure 6-8. Freeze–thaw damage.
Source: Courtesy of Vector Corrosion Technologies.

spalls may simply be repaired, but active spalls are warnings of a greater ongoing problem, possibly a dangerous problem, or even affecting the integrity of the structure. Figure 6-11 depicts a concrete spall caused by corrosion of reinforcement and an incipient spall above.

Figure 6-9. Popout at aggregate caused by a cement-aggregate reaction.
Source: Courtesy of Raths, Raths, & Johnson, Inc.

Figure 6-10. Concrete scaling.
Source: Courtesy of Raths, Raths, & Johnson.

Figure 6-11. Spall and insipient spall in a corroded reinforcement.
Source: Courtesy of Raths, Raths, & Johnson.

Stratification: Segregation of the constituents of a concrete mix caused by overwet or overvibrated concrete, resulting in the formation of horizontal layers with increasingly lighter material toward the top.

Sulfate Attack: Reaction between sulfate solutions (resulting from, for instance, alkali magnesium and sodium sulfates in groundwater and seawater) and hardened cement paste (the sulfate reacting with the calcium aluminate in the cement paste). This reaction can lead to an expansion and cracking of the concrete.

Concrete may crack because of internal expansion resulting from the reaction of sulfates in solution with aluminate hydrates of the cement. If the amount of aluminate hydrate is sufficiently large, such cracking may take place only because of an interaction with moisture from the environment if the source of the sulfates were in the cement or the aggregate. This, however, is not the normal condition; usually, the source of the sulfates comes from the environment (Mather 1968).

Unsound Cement: Concrete that has undergone deterioration or disintegration during service exposure. The presence of compounds in the cement results in appreciable expansion under conditions of restraint, causing disruption of the hardened cement paste. Such expansion may take place because of delayed or slow hydration of free lime, magnesia, and calcium sulfate.

Concrete may crack because of internal expansion caused by the reaction of moisture with anhydrate calcium oxide or magnesium oxide that was introduced into the concrete as part of the cement (Mather 1968).

6.1.7 Factors That Influence Concrete Behavior

6.1.7.1 Causes of Concrete Deterioration. Concrete deterioration can occur because of alkali–aggregate reactions, unsound cement, contaminated water and aggregates, sulfate attack, freezing and thawing, fatigue, poor detailing, and damage from accidents and overloading. Some of these causes are normally related to deterioration of the reinforcing steel and the prestressing steel embedded in the concrete.

Accident and Overloading Damage: Concrete is a brittle material, and when struck by a vehicle, for example, a brittle fracture can result. In general, exposing concrete building elements to vehicle traffic should be avoided, or the elements should be protected against impact damage. When this is not done, damage could be the result. Examples of accident damage include columns and walls unprotected by guard rails in garages and loading areas. Overloading of a structure can cause fractures or even partial or full collapses. Overloading can be from changing of occupancy loadings, unauthorized vehicles, or even ponding caused by poor roof drainage.

Alkali–Aggregate Reaction (AAR): Reaction in concrete between alkalis from the cement in the pore water of the concrete and reactive silica in the aggregate is termed "alkali–aggregate reaction" (Struble and Diamond 1986). This reaction can be accelerated by surface wetting and exposure to high-humidity environments. Cracking and an occasionally significant weakening of the concrete can result from this reaction. Three types of alkali–aggregate reactions are believed to exist: alkali–silica, alkali–silicate, and alkali–carbonate. The alkali–silica reaction is the most common (Hobbs 1988). The alkali–silica reaction can cause substantial reductions in the engineering properties of concrete (Swamy and Al-Asali 1986). At 0.1% expansion, compressive strength decreases by about 12%, loss of flexural strength can be as much as 50%, and the elastic modulus reduces by approximately 20%. Sealing or coating the surface to limit or prevent moisture from permeating the concrete slows down the reaction. Few structures have been replaced because of AAR, because when it occurs, it can be mitigated by treating it with a lithium solution.

Cement Soundness: Portland cement as a hydrated paste is the most common binder of most concretes. This binder governs many of the properties of concrete. Cement is available in several types, with and without air-entraining agents, to meet various construction requirements (such as early high strength) and for compatibility with certain types of aggregates. Cement can become contaminated, usually when it is transported or stored. Contamination by organic materials can result in a loss of the required properties of the concrete. Specifications for cement manufacture specify limitations in fineness and specific surfaces. Cements with a specific surface that is too low may produce concrete with poor workability and excessive bleeding. Decreased fineness increases water requirements, whereas a greater fineness improves early strength development but reduces resistance to freezing and thawing. Because of differences in raw materials, cements with the same fineness and chemical analyses can have differences in strength, heat of hydration, production of laitance, bleeding tendency, and durability (U.S. Department of the Interior 1981). The alkalis in the cement can react with minerals in the aggregates, resulting in cracking and expansion. Sulfates in groundwater and soil, against which concrete is placed, can also react with chemicals in the cement, resulting in the disintegration of the concrete. Cement must be selected in such a way that it is compatible with site conditions.

Contaminated Water and Aggregates: Usual rule governing mixing water is that if it is potable, that is to say, meets the requirements for drinking water, it is satisfactory for making concrete. This is not an entirely satisfactory definition because water containing sugar or citrate flavoring are fit to drink but not for mixing concrete (McCoy 1966). Mixing water can be contaminated by impurities such as sodium and potassium carbonates and bicarbonates, sodium chloride, lithium bromide, sodium sulfate, calcium and magnesium bicarbonates, calcium chloride, iron salts, sodium iodate, phosphate arsenate, borate, sodium sulfide, hydrochloric and sulfuric acids, sodium hydroxide, and salt and suspended particles. It can also be contaminated by algae and other organic materials. Minute concentrations are tolerable, and limits on the concentration of different impurities are available (McCoy 1966). Aggregates for concrete should usually be inert. However, aggregates may contain minerals that react unfavorably with the chemicals in the cement paste. The reactive components of aggregates include calcium, magnesium, silica, and iron oxides; ferrous sulfides; glasses; calcium sulfate; zeolites; clay minerals; and dolomitic limestone (Hansen 1966). As with unsound cement, some contaminated aggregates may be perfectly acceptable for specific uses, construction methods, and with the other components of the concrete mix, but the opposite may be true for different conditions.

Cracking, Scaling, Spalling, and Delamination: Crack is an incomplete separation of concrete into one or more parts of variable depth but visible at the surface. Plastic shrinkage cracks result from rapid drying of the concrete in its plastic state and are often random in pattern. Map cracking is a closely spaced network of cracks usually resulting from sulfate attack, frost attack, and chemical reactions between the mineral aggregates and the cement paste. Drying shrinkage cracks result from the drying of restrained concrete after it has hardened and often form a pattern. Settlement cracks can result from settlement of the formwork as well as from settlement of the foundation. Structural cracks result from unreinforced or overstressed tension areas in the concrete. Corrosion-induced cracks are caused by corrosion of reinforcing steel in the concrete (Section 6.1.13, Table 6-4, Manning 1985). Most surface cracks are primarily caused by concrete volume changes resulting from poor construction practices such as the use of concrete that is too wet, placed too slowly, poorly finished, and improperly cured (Peterson 1977).

Scaling is the flaking away or disintegration of the cement matrix surface of concrete that exposes and eventually loosens coarse aggregate. Although a weak surface layer of mortar may result from poor finishing and curing practice, it is usually indicative of inadequate air entrainment (Section 6.1.13, Table 6-4) (Manning 1985). Scaling is not directly related to traffic (Peterson 1977). It is often caused by freeze–thaw action.

Spalling is most often caused by corrosion of reinforcing steel in the concrete and by impact, which causes local disintegration of a portion of the concrete surface (sometimes referred to as potholes).

Delamination usually occurs over subsurface areas of corroding reinforcing steel, resulting in a plane of separation within the concrete immediately over the plane of the reinforcing steel. If planes of reinforcing steel occur at several depths and the steel is corroding at each depth, then there may be more than one plane of delamination.

Degradation: Concrete can degrade when it comes into contact with both organic and inorganic acids. Common organic materials include acetic acid, oxalic and dry carbonic acids, carbonic acid in water, lactic and tannic acid, and vegetable oils. Common inorganic acids include hydrochloric, sulfuric, nitric, and phosphoric acid. Degradation can also be the result of an environmental attack by wind and rain.

Electrical Resistivity: Resistivity is the reciprocal of conductivity and measures the resistance to flow through an electrolyte (such as soil, sea/saltwater, or concrete) and is affected by the ion content of the electrolyte. The fewer the ions, the greater the resistivity, the less current for a given cell voltage, and the lower the rate of corrosion.

Fatigue: Damage of reinforced concrete is dependent on the fatigue life of both the reinforcing steel and the concrete. Like all fatigue damage, the number of cycles of stress and the stress range are the most important parameters affecting fatigue damage. The failure of concrete under repeated loading results in progressive microcracking of the concrete. Because localized plastic deformations to blunt microcracks, such as those occurring in metals, do not occur in concrete, concrete has no endurance limit similar to that for steel (Hawkins and Shah 1982). The fatigue strength of concrete submerged in water is different from that of concrete in an air environment. Water trapped in the opening and closing of the cracks causes hydraulic fracturing. Submerged concrete has a shorter fatigue life than unsubmerged concrete. Fatigue of concrete can be of several different types or combinations of types. These include shear fatigue, tension fatigue, and bond fatigue. Under static loads, microcracking associated with these failure mechanisms may be insignificant, but the repetition and range of stresses can result in a breakdown of resistance. Research on concrete indicates that fatigue damage of concrete may be more prevalent than realized (Lenschow 1982). Also, a higher degree of utilization of the capacity of materials, which increases stress variations and brings the service stress range closer to the failure stress, will result in a greater amount of concrete fatigue damage. Much of the spalling of concrete from the underside of concrete parking garage decks and reinforced concrete beams and girders subjected to cyclical stresses is suspected to be fatigue related.

Fire: Fire can damage reinforced concrete in several ways. Surface defects such as spalling and cracking can occur as a result of internal pore pressure buildup, induced thermal stresses caused by a severe temperature gradient, or from differences in the thermal expansion characteristics of two, unlike materials. Over time, with increasing fire exposure, both concrete and steel will reach temperatures that result in loss of strength. Cement paste decomposes and progressively begins to lose strength at about 300 °C (500 °F).

Freezing and Thawing: Disintegration and cracking of concrete can be caused by the disruptive action of freezing and thawing. Dry concrete, with or without entrained air, sustains no damaging effects from freezing and thawing. It is the water that is available in the capillary system that freezes and expands, which causes concrete disintegration. In general, the problem is controlled by air entrainment that provides relief for pressures developed by free water as it freezes and expands.

Sulfate Attack: When sulfates are in groundwater (or mixing water) or in soils against which concrete will be in contact, they can react with the cement paste in concrete and result in disintegration of the concrete. The problem is fairly common in arid regions and in exposures to seawater (Mehta and Polivka 1975, Reading 1975). Sulfates of sodium and magnesium can react with the cementing constituents of the concrete. Depending on the composition of the cement, these chemical reactions may be accompanied by large volume changes in the concrete.

Poor Detailing: Poor reinforcement detailing can lead to a reduction in the capacity of the concrete members. The lack of proper reinforcement within concrete members can lead to excessive deflections and cracking.

6.1.7.2 Causes of Reinforcing Steel Deterioration. Reinforcing steel is a component of reinforced concrete. The primary cause of deterioration of reinforcing steel is corrosion. A secondary cause is fatigue cracking. Both these types of deterioration trigger deterioration in the surrounding concrete. Although both materials have separate deterioration mechanisms, the deterioration of one affects the other.

Corrosion: When reinforcing steel is surrounded by concrete, a passive environment for the steel is provided. Fresh concrete is highly alkaline, with a pH usually in the range of 12 to 13. In this alkaline environment, the thin film of iron oxide that is usually present at the surface of the metal is stabilized into a protective passive film (Borgard et al. 1990). This passive film protects the reinforcing steel against corrosion. In addition, well-cured concrete of low water–cement ratio has low permeability, which minimizes the penetration of corrosion-inducing factors, especially oxygen, chlorides, and carbon dioxide. Low permeability also increases the electrical resistivity of the concrete, which assists in reducing the rate of corrosion by retarding the flow of electrical currents that accompany electrochemical corrosion (Verbeck 1975).

Corrosion of reinforcing steel is the electrochemical degradation of the steel. It occurs in the presence of oxygen and moisture whenever the protective passive film is being destroyed, most often by either carbonization or the presence of more than the threshold concentration of chloride ions at the steel surface. As a result, the alkalinity of the concrete (pH) is reduced, and the steel loses its protective film. Chloride ions can also penetrate the protective passive film, with an accompanying shift in steel electric potential, and subject the steel to the action of oxygen and moisture and resulting corrosion (Locke 1986). The higher the alkalinity or pH of the concrete, the higher the threshold concentration of chloride ions it takes to initiate corrosion.

Once the protective oxide film has been impaired, and in the presence of oxygen and moisture, corrosion of the steel will proceed. There are two mechanisms of steel corrosion: one electrochemical/physical and the other galvanic/physical. In marine environments or where salt is used as a deicing agent, as in parking garages or areas open to vehicular traffic, chlorides will permeate the concrete by absorption, capillarity, or diffusion through cracks (Mehta 1977). Chloride ions depassivate steel, which results in electrochemical corrosion cells and, with oxygen and moisture, accelerates attack on the reinforcing steel. Corrosion products from the attacked steel create expansive forces (physical) within the concrete, causing it to delaminate, crack and spall, or crumble. As the steel is further exposed to oxygen, CO_2, and Cl^-, it corrodes more and increases deterioration of the concrete in a progressive manner.

Likewise, an imbalance of temperature, moisture, oxygen concentration, atmospheric pressure, stray current, or other variables produces corrosion cells. The resulting products of corrosion cause deterioration of the concrete. Reinforced concrete is most vulnerable to corrosion when reinforcing steel embedded in concrete is exposed to seawater.

Detection of corrosion or corrosion potential within concrete can be performed by numerous means, of which corrosion potential mapping (Figure 6-12), half-cell potential testing (Figure 6-13), and corrosion rate testing (Figure 6-14) are a few.

Fatigue: Fatigue failures of reinforcing steel in tension can usually be attributed to the magnitude of stress range in the reinforcement and stress risers provided by deformation lugs (Helgason et al. 1976). There are other contributing factors, for example, stress concentration at critical points such as at bar anchorages, splices, and bar bends and friction (fretting) between the reinforcing steel and the concrete at a cracked section (Naaman 1989). Current design specifications provided by ACI 343 and AASHTO address fatigue stress limits in reinforcing steel under the Strength Design Method and under Load and Resistance Factor Design. Both specifications limit the stress range of straight tensile reinforcement. It is concluded that the current design specifications are conservative, at least for straight tensile reinforcement.

Most fatigue failures in reinforcing steel can be associated with the stress concentration at critical points. These include butt-welded reinforcement (Walls et al. 1965), splice couplings and sleeves (Bennett

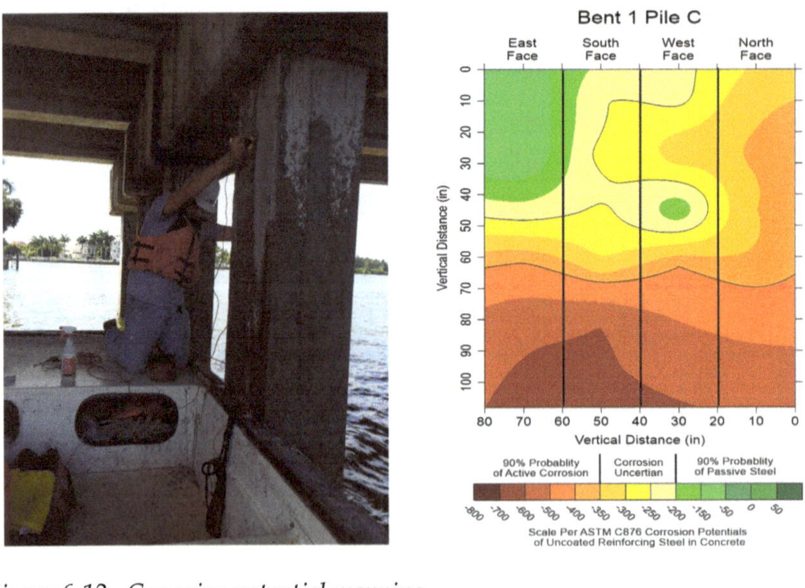

Figure 6-12. Corrosion potential mapping.
Source: Courtesy of Vector Corrosion Technologies.

Figure 6-13. Half-cell potential testing.
Source: Courtesy of Vector Corrosion Technologies.

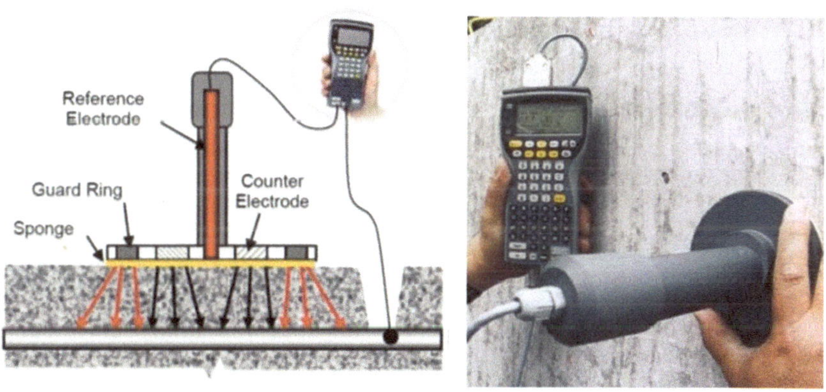

Figure 6-14. Corrosion rate testing.
Source: Courtesy of Vector Corrosion Technologies.

1982, Kordina and Quast 1982, Tassi and Magyari 1982), areas where notch effects of the deformations are present, and the ratio of curvature of bent bars (Nurnberger 1982).

Because allowable design stresses for fatigue of reinforcing steel are, in general, conservative, fatigue failure can usually be associated with corrosion of the reinforcing steel rather than design stresses being exceeded for unaffected rebar. High stresses are present at areas of corrosion, making these areas more vulnerable to fatigue.

6.1.7.3 Causes of Prestressing Steel Deterioration. Like reinforcing steel, prestressing steel deterioration can result from corrosion and fatigue. Because of the high stresses associated with prestressing steel, hydrogen embrittlement and stress corrosion are also possible causes of deterioration. Pretensioned steel in precast concrete, like reinforcing steel, is protected against corrosion by being in direct contact with the high alkalinity concrete. When prestressing steel is posttensioned, it is usually encased in paper, plastic, or corrugated light gauge sheathing or conduits. When these conduits are grouted full, the prestressing steel is considered bonded; otherwise, it is considered unbonded. For unbonded posttensioning, the lack of grouting of these conduits permits oxygen and moisture to be present adjacent to the prestressing steel surfaces, which can result in corrosion. The sheathing or conduits used in early unbounded posttensioning systems, which used paper wrap or loose-fitting plastic conduits with grease-coated prestressing steel, are more susceptible to corrosion than later, tighter-fitting conduits where the prestressing steel is encapsulated in extruded plastic. For bonded posttensioning steel, inadequate or poor grouting can lead to steel deterioration, especially when excess bleed water is present in the grout. Excess water can contribute to grout consolidation and segregation, which can lead to voids and to prestressing steel corrosion.

Anchor: Steel element is either cast into concrete or postinstalled into a hardened concrete member and used to transmit applied loads to the concrete.

Anchorage Device: In posttensioned members, it is the hardware used to transfer force from prestressed reinforcement to the concrete.

Corrosion: Corrosion of prestressing steel appears to be more highly sensitive to permeable concrete associated with a high water-to-cementitious materials ratio, honeycombing, and large bleed holes (Gerwick 1975). Also, stray electric current can affect the corrosion process for prestressing steel, at least for prestressed concrete exposed to seawater, more than for reinforcing steel (Cornet and Bresler 1980). Corrosion of prestressing steel is also usually associated with catastrophic failure because of the high magnitude of force in the strands and the lack of redundancy in typical design (Rothman and Price 1986). Anchorages and couplings for prestressing steel usually have less protective concrete cover and are accordingly more susceptible to corrosion.

Detection of corrosion or corrosion potential within concrete can be performed by numerous means, of which corrosion potential mapping magnetic flux-cable break detection (Figure 6-15) and posttensioning humidity/corrosion potential testing (Figure 6-16) are a few.

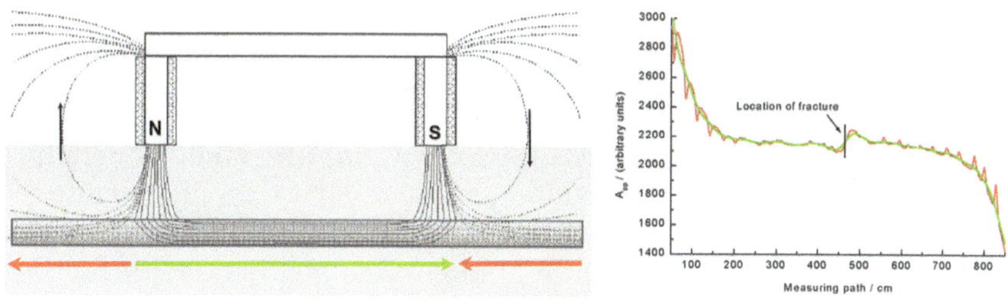

Figure 6-15. Magnetic flux-cable break detection.
Source: Courtesy of Vector Corrosion Technologies.

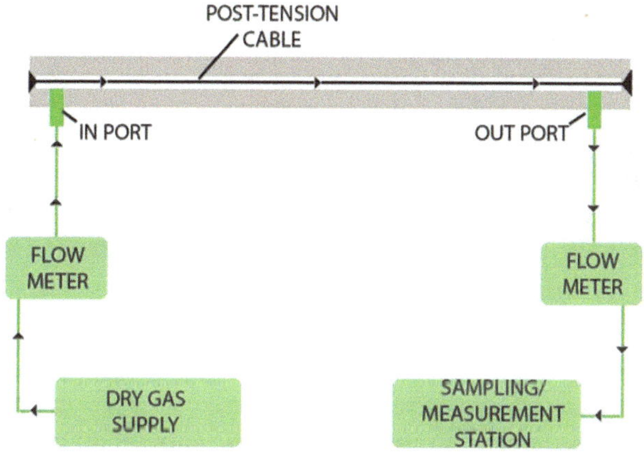

Figure 6-16. Posttentioning humidity/corrosion potential testing.
Source: Courtesy of Vector Corrosion Technologies.

Fatigue: This discussion on fatigue of reinforcing steel applies directly to prestressing steel. Points of critical stress for prestressed members occur at changes in the direction of the prestressing steel and the anchorages. In partially prestressed concrete beams, the concrete section is, in general, uncracked under dead load, with cracking initiating under the first application of live load. Subsequent applications of live loads will lead to cracks opening at the decompression stress, which is lower than the cracking stress.

Once cracking in the concrete initiates, the location of the neutral axis shifts upward or downward, depending on the flexural direction of the structural component. These repetitive changes in stress create fatigue damage in the corresponding materials, reduce bond properties at the interface between the prestressing steel and the concrete, and lead to substantial increases in crack widths and deflections under service loads (Naaman 1982). Consequently, partially prestressed concrete members, in general, have shorter fatigue lives than fully prestressed concrete members.

Hydrogen Embrittlement: Hydrogen embrittlement is a failure mechanism of metal resulting from the entry of hydrogen into the metal, which reduces its ability to deform plastically (Sheinker and Wood 1972). It is quite often identified as one of the two mechanisms of stress corrosion (the other mechanism of stress corrosion is identified as the active path corrosion type) but is usually treated as a separate subject. For either hydrogen embrittlement or stress corrosion cracking to occur, the following conditions are necessary (Sheinker and Wood 1972):

Susceptible Metal: Metal may be susceptible to hydrogen embrittlement or stress corrosion cracking in only a few specific environments.
Specific Environment: Particular environment may induce cracking in only certain metals.
Tensile Stress: Tensile stress usually must exceed a certain level depending on the particular metal–environmental couple.

ASTM A143/A143M defines hydrogen embrittlement as the loss or partial loss of ductility in steel often associated with strain aging, which proceeds slowly at room temperatures or more rapidly as the aging temperature is raised. Accordingly, galvanized prestressing steel is more susceptible to hydrogen embrittlement than ungalvanized prestressing steel because of the galvanizing temperature of approximately 850 °F (455 °C). Hydrogen embrittlement can result from atomic hydrogen being absorbed by the steel during the steel manufacturing and fabrication processes. It can also result from hydrogen introduced during service or environmental exposure, such as during welding (Raymond 1988).

One reason for separating the phenomenon of hydrogen embrittlement from stress corrosion is that the two types of cracking failures respond differently to environmental variables (Fontana and Greene 1978). For example, cathodic protection is an effective means of preventing stress corrosion cracking, but it rapidly accelerates hydrogen embrittlement effects.

1. **Inadequate Grouting**: Prestressing steel for posttensioned structures is installed in conduits to prevent bonding with the concrete when it is placed. After posttensioning, these conduits are either partially filled with an organic substance, such as grease, more fully filled with extruded plastic to prevent bonding, or filled with a cement grout to provide a passive protective environment for the prestressing steel and to bond the prestressing

steel to the concrete. Nongrouted conduits are termed "unbonded," and cement-grouted conduits are termed "bonded conduits." It is noted that both types of construction have a similar problem with regard to corrosion. When the conduits are filled with a portland cement grout, because of excessive consolidation, inadequate filling of the conduit can occur. The resulting voids permit water leakage and oxygen into the ducts, causing corrosion. This water can contain impurities leading to more serious corrosion (Griess and Naus 1980), or the water can freeze with the resultant expansion cracking the concrete, usually at locations of minimum cover. In addition, the mortar can segregate into different strata or layers, each having different properties. Usually, the higher grout strata are more water-soluble and have higher water content and a lower pH. Therefore, adjacent sections of the prestressing steel can be in different concrete environments, which can result in the development of localized corrosion cells.

2. **Stress Corrosion**: Stress corrosion cracking (SCC) is the cracking induced from the combined influence of tensile stress and a corrosive environment. The impact of SCC on a material usually falls between dry cracking and the fatigue threshold of that material. The required tensile stresses may be in the form of directly applied stresses or in the form of residual stresses. Corrosion plays an important part in the initiation of stress corrosion cracks. When a corrosion pit first forms, it acts as a stress raiser. There is a porous cap to the corrosion pit, which impedes exchange between the corrodent within the pit and the bulk corrosion outside the pit. The essential feature of the corrosion pit is to provide a mechanism for altering the solution chemistry locally to one favorable for stress corrosion cracking. With the passage of time, cracks emanate from the pit downward into the metal. These cracks can grow to lengths sufficient for the remaining metal to fail in a purely mechanical brittle fracture (Brown 1972). As noted under Hydrogen Embrittlement, this mechanism is termed "active path corrosion."

The characteristics of stress corrosion cracking may be summarized as follows (Brown 1972):

- Tensile stress is required. This stress may be supplied by prestressing, service loads, cold work, mismatch in fit-up, heat treatment, and by the wedging action of corrosion products or because of the steel manufacturing process itself.
- In general, only a few chemical species in the environment are effective in causing stress corrosion cracking of a given alloy.

- Species responsible for stress corrosion cracking, in general, need not be present either in large quantities or in high concentrations.
- With some alloy/corrodent combinations, such as titanium and crystalline sodium chloride or austenitic stainless steel and chloride solutions, temperatures substantially above room temperature may be required to activate the processes essential to stress corrosion cracking.
- Alloy is usually almost inert to the environment, which causes stress corrosion cracking.
- Stress corrosion cracks are always macroscopically brittle in appearance, even in alloys that are very tough in purely mechanical fracture tests. (Shear lips may occur in conjunction with stress corrosion cracks, but these shear lips are not part of the stress corrosion process. As a corollary, there does not appear to be a stress corrosion fracture analog to the full shear slant fracture of purely mechanical origin.)
- Microscopically, the fracture mode in stress corrosion cracking is usually different from the fracture mode in-plane strain fractures of the same alloy.
- There appears to be threshold stress below which stress corrosion does not occur, at least in some systems.

6.1.8 Considerations with Concrete in Historic Construction

Concrete has been used for utilitarian, ornamental, and monumental structures since ancient times. Although often considered to be permanent, it is, like all materials, subject to deterioration. Historic concrete is usually found in structures of historical, architectural, or engineering significance. In the United States, concrete has been a popular construction material since the late nineteenth century, and concrete has been recognized by various governmental and private organizations as a historic material.

Preservation of historic concrete requires a thorough understanding of the causes and types of deterioration, as well as of repair and replacement materials and methods. Adequate time needs to be allowed during the planning phase of a project to provide for trial repairs and mock-ups so as to evaluate the effectiveness and aesthetics of the repairs. Careful design is essential, and, as with other preservation efforts, the skill of those performing the work is critical to the success of the repairs. The techniques and materials now available can extend the life of such structures and help ensure their preservation well into the future.

Materials and workmanship used in the construction of historic concrete structures can be potential sources of problems, such as poor consolidation of the concrete during its placement. The type of cementitious material, type of aggregate, and others used usually need to be determined. The significance of a historic concrete building or structure and the costs to maintain and rehabilitate all help determine if repair is needed and will be undertaken. Determining the causes of deterioration is also central to the development of a conservation and repair plan. Allowing sufficient time during the planning phase to analyze the concrete, develop mixes, and provide time for adequate aging of mock-ups for matching to the original concrete is imperative.

Determining the original construction techniques, some no longer known or practiced, and previous repair work performed on historic concrete is also important. Information can be obtained through document research, visual survey, inspection, laboratory studies, and field testing. Any proposed preservation plan should involve professionals experienced in historic concrete restoration who are most capable of developing proper repair approaches.

Documents related to past repairs, such as archival photographs, can provide a valuable source of information about original construction as well as past repairs. Selection and design of proper repair materials is a critical component of the repair project. Selected materials need to be compatible with the original concrete. Repairs should retain as much of the original material as possible, while allowing for the removal of an adequate amount of deteriorated concrete to provide a sound substrate for a durable repair, while matching existing concrete as closely as possible. High levels of craftsmanship are required for matching the original historic concrete. Achieving compatibility between new repair and restoration work with the original concrete is always difficult.

6.1.9 Connectors and Connections

A condition assessment of connections between building framing members includes determination of the material used, whether it is the proper type, the condition it is in, and whether it is properly designed and constructed. The assessment may require both visual and nondestructive or laboratory testing by qualified personnel. Because steel embedment connections or postinstalled anchorages are often used, a review of existing contract documents and shop drawings can be useful.

Connections in cast-in-place structural concrete (unreinforced, reinforced, pretensioned, posttensioned, or combinations thereof) are usually integral with members; consequently, there are few if any separate connections to be assessed.

Precast structural concrete construction and prestressed (pretensioned or posttensioned or combinations thereof) attachments between different adjacent members are always of nonconcrete materials (e.g., different types of bearings or embedded or attached steel shapes). Reference the appropriate material section of this standard for guidelines on the assessment of such connections.

Design criteria for precast connections are given in the PCI Design Handbook and ACI 318 (Chapter 17). Design for steel connectors is governed by the AISC Steel Construction Manual.

6.1.10 Terminology Related to Connector Characteristics and Properties

Concrete connectors are the same as used for metals, wood, and other materials. Reference is made to Section 6.2.1 for the appropriate properties.

6.1.11 Physical and Mechanical Properties of Connectors and Connections

For cast-in-place structural concrete (unreinforced, reinforced, pretensioned, or combinations thereof), the connections are primarily integral with the members, and there are few if any separate connections to be assessed. For precast concrete (unreinforced, reinforced, pretensioned, posttensioned, or combinations thereof), the connections are always of different materials (e.g., different types of bearings or embedded or attached steel shapes that are attached by connectors), and reference is made to the appropriate material section in this standard for assessment of the appropriate physical conditions. In reinforced concrete, an extension of the reinforcement, either straight or bent, is used to transfer the force in the reinforcement to support. In pretensioned and posttensioned concrete, a separate device is used for the same purpose. The following connector considerations should be considered:

1. **Corrosion Protection**: Disintegration or deterioration of reinforcement by electrolysis or chemical attack needs to be prevented.
2. **Cover**: The least distance between the surface of the reinforcement and the outer surface of the concrete.
3. **Cross-Sectional Properties**: Actual dimensions and other geometric properties of reinforcement.
4. **Exposure**: Condition where reinforcement has been visibly exposed by the loss of its protective cover.
5. **Dimensional Location**: Spacing, position, and length of reinforcement.
6. **Shape**: Physical configuration of reinforcement; its bent shape, which may be either two- or three-dimensional.
7. **Ductility**: Mechanical property that measures the ability to deform before cracking and being permanently deformed or damaged.

8. **Elongation**: Increase in gauge length of a tension test specimen measured after a fracture as a percentage of the original length (a measure of durability).
9. **Fatigue Properties**: Mechanical properties, such as cyclic counting, endurance limit, *e*-*N* curve, hardness, and *S*–*N* curve that characterize the resistance of metals to fatigue crack initiation and propagation under low-amplitude high-frequency stresses or strains.
10. **Modulus of Elasticity**: Ratio of normal stress to the corresponding strain in the linear elastic segment of the stress–strain curve.
11. **Tensile Strength**: Stress calculated from the maximum load sustained by a specimen during a tension tensile test divided by the cross-sectional area of the specimen.
12. **Yield Strength**: Stress at which a significant increase in strain occurs without an increase in stress, as defined by the appropriate standard specification for the material.

6.1.12 Mechanism of Connection and Connector Deterioration

Connectors are typically metal and are susceptible to the same methods of deterioration as metals, as discussed in Section 6.2.7.

6.1.13 Test Methods

6.1.13.1 Test Methods (Preliminary Assessment Chart). The relative value to the assessment professional of visual observations, nondestructive testing (NDT), and destructive field or lab testing is depicted in Table 6-1. This chart is a quick reference for the applicability of certain test methods to determine concrete and concrete connection conditions and properties. Subsequent tables list the capabilities and limitations of various test methods in greater detail, along with a list of references that have additional information. Condition assessment of structures is dynamic, and the investigator should recognize that some technologies listed may become unavailable as devices are removed from the market or replaced by newer technology.

New test procedures, methodologies, and equipment are developed on an ongoing basis, but most are not commercially available or have gained wide acceptance and should, therefore, be used with a sense of caution. It may be necessary to use a combination of methods to satisfactorily predict the strength or quality of concrete. References cited in Section 6.1, unless otherwise noted, are included in Appendixes B1 and B3. Additional references are to be found in Appendixes B2 and B4.

SAMPLE SIZE—Several ASTM standards exist to guide an investigator in determining sample size if a statistically significant sample is desired. Often a statistically significant sample is not available or economically viable. In this case, it is desired to select a sample without bias. This sample could also be a random sample, but not necessarily, as purely random samples in a failure investigation are often not practical given the constraints of the existing structure. As sample size and distribution are determined, care should be exercised to obtain samples from damaged and undamaged and questionable and unquestionable portions of the structure if applicable.

ASTM E122, *Standard Practice for Calculating Sample Size to Estimate, with a Specified Tolerable Error, the Average for a Characteristic of a Lot or Process* (ASTM 2017), can be followed to determine the number of samples required for a desired level of accuracy. The number of samples will depend on

- Error E that can be tolerated in the estimate of the measured value,
- Preferred probability, or confidence level, that the measured value is within the limit of E, and
- Estimate of the variation, σ, of the measured value from one unit to another within the population.

Assuming that the measured result is normally distributed and that the population is large, the following equation, based on an assumed 95% confidence interval, can be used to calculate the number of samples, n, for testing:

$$n = \left(\frac{1.96\sigma}{E}\right)^2 \tag{6-1}$$

where

n = Number of samples,
σ = Standard deviation of the measured results, and
E = Allowable error in the measured results.

The factor 1.96, based on a desired 95% confidence level, gives a 0.050 probability that the actual value is not within plus or minus the allowable error, E, of the measured property.

Deterministic sampling describes a situation where a set number of samples are available, or sampling is constrained by any of the factors discussed herein, and only a set number of samples can be taken (access, effects on the element under consideration, budget, etc.). This sampling protocol requires engineering judgment to ensure that a representative estimate of the characteristic under consideration is obtained.

Table 6-1. Relative Assessment Value for Concrete and Connection Conditions and Properties.

Factor	Method: In situ visual inspection	Petrographical examination	Ultrasonic pulse velocity	Pachometer (metal detectors)	Ground-penetrating radar	Infrared thermography	Moisture measurements	Impulse response	Half-cell/linear polarization	Sampling and testing	pH meter	Destructive openings	Radiography
Material properties concrete and steel										Yes			
Measure moisture content						Limited	Yes						
Locate deterioration	Limited		Yes					Yes				Yes	
Quantify deterioration	Limited	Yes							Yes		Yes	Limited	
Assess strength			Yes							Yes			
Determine modulus of elasticity										Yes			
Identify hidden construction defects					Limited			Limited				Yes	
Locate/assess reinforcement		Yes		Yes	Yes							Yes	Yes
Locate/assess connections	Limited			Yes						Yes		Yes	Yes

Some guidance is provided within concrete codes, standards, and guides regarding concrete sampling. For compressive strength, at least three cores should be taken at each location in the structure, and the strength value shall be the average of the three cores. The selection of test methods and the number of tests and their locations depend on

- Variation in material properties within the structure;
- Variations in exposure, loading, and use;
- Critical locations, such as connections and lateral loading transfer points;
- Probable error in a test result; and
- Extent of the structure over which a property is measured, full depth versus partial depth.

6.1.13.2 Test Methods (Comprehensive Assessment Charts). Testing of specimens from the members of a structure to determine chemical and physical properties can provide useful quantitative data. Although removal of specimens may be destructive, careful selection and handling can ensure that the usefulness of the member is not impaired, and the damaged areas can be satisfactorily repaired. Tests on specimens provide specific, quantitative information without a reliance on interpretation or analytical results. Consideration should be given to the physical testing of specimens over nondestructive testing methods to determine similar information. It is also advantageous to use the results of physical tests in specimens from members as base data to corroborate nondestructive testing results. Most nondestructive methods require highly qualified and skilled personnel, specialized equipment, and considerable analysis and interpretation of data to provide qualitative and quantitative information.

Available test methods are cited in Tables 6-2 through 6-6. The tables are not intended to provide the reader with information and details on how to perform and interpret tests, but rather to provide information on test methods that are available, what they can accomplish, advantages and limitations of the tests, and a brief statement about qualifications of testing personnel. This information is not exhaustive and is constantly subject to change. Advantages, limitations, and costs can change with time. Accordingly, the references are the most important part of these tables, and care has been taken to provide the most current and in-depth information. A perusal of any specific reference will usually lead to an extensive bibliography on the subject. The supplemental references in Appendixes B2 and B4 contain information that can be pursued for additional knowledge.

Table 6-2. Test Methods for Determining Chemical and Physical Properties of Concrete.

Property	Purpose of test	Principle of operation	Advantages	Limitations
Air–void content	Determination of the proportional volume of air voids in concrete expressed as volume percent of the hardened concrete includes both "entrapped" and "entrained" air.	Measurement of the air void system in hardened concrete by prescribed microscopical procedures on sawed and ground sections from specimens taken from the hardened concrete.	Common tests providing useful information.	Difficulties may arise in preparing a satisfactory surface from the specimen for analysis.
Chloride content	Determination of chloride content.	Samples are pulverized, weighed, mixed with water, treated chemically, heated, filtered, and volume-passed-tested for chloride ion content.	Common tests providing useful information	Multiple samples are helpful to determine the chloride ion profile in a concrete member.
Compressive strength	Determination of compressive strength from drilled cores and portions of beams broken in flexure.	Prepared specimens are tested to fracture in a testing machine.	A common test providing useful information; low in cost.	Correction factors for specimens with L/D less than 1.8 are approximate and subject to variability.

(Continued)

Table 6-2. (*Continued*)

Property	Purpose of test	Principle of operation	Advantages	Limitations
Flexure strength	Determination of flexural strength (modulus of rupture) and, indirectly, tensile strength of the specimen.	Loading of a standard beam test specimen until a fracture occurs.	Provides useful information.	Precision of test methods has not been established. Not a common test because of the size of the required specimen.
Fundamental frequencies	Determination of significant change in the dynamic Young's modulus of elasticity, dynamic modulus of rigidity, and dynamic Poisson's ratio.	Specimens are forced to vibrate at varying frequencies. A well-defined peak of response is the fundamental frequency.	Assessment of the uniformity of concrete	Costly. Results cannot be used for design.
Thickness	Determination of the thickness of concrete slabs and walls for compliance with requirements.	Measurement of drilled core length by standard procedures.	A common test low in cost	Problem with cores taken from concrete placed against earth because of a great number of projections and voids.

(*Continued*)

Table 6-2. (*Continued*)

Property	Purpose of test	Principle of operation	Advantages	Limitations
Petrographic analysis	Determination of acidity, alkali–carbonate reaction; alkali–silica reaction chemical composition; contaminated mixing water and aggregates; quality, uniformity, and soundness of aggregates; water–cement ratio and frozen aggregates.	Prescribed microscopic examination on sawn and ground sections from specimens taken from hardened concrete.	Common tests providing much useful information.	Difficulties may arise in preparing a satisfactory surface from the specimen for analysis. Costly. Results are greatly enhanced if results can be compared to the results of other specimens from the structure.
Potential for volumetric change	Measurement of length change permits assessment of the potential for volumetric expansion or contraction because of causes other than applied force or temperature.	Determination of length changes after both air and water storage.	Provides useful information.	Costly.

(*Continued*)

Table 6-2. (*Continued*)

Property	Purpose of test	Principle of operation	Advantages	Limitations
Portland cement content	Determination of portland cement content in percentage by mass.	Crushed and graded samples are heated and cooled, subjected to a soluble silica subprocedure, or a calcium oxide procedure, filtration, and measurement of the silica content of concrete.	Provides very useful information.	Costly. Precision is dependent on the composition of concrete and on the subprocedure used.
Resistance to freezing and thawing	Determination of the resistance of concrete specimens to rapidly repeated cycles of freezing and thawing.	Test specimens under controlled moisture conditions are subjected to standard cycles of freezing and thawing in water or by rapid freezing in air and thawing in water. Specimens are measured for length and mass before and after tests, and percentage change recorded as well as a calculated durability factor.	Provides useful information on durability and service life.	Precision of results is dependent on the number of tests.

(*Continued*)

Table 6-2. *(Continued)*

Property	Purpose of test	Principle of operation	Advantages	Limitations
Specific gravity (density), percent absorption, and percent voids.	Determination of specific gravity, percent absorption, and voids in hardened concrete.	Specimens are oven-dried, mass-determined, immersed in water for 48 h, surface-dried, and surface-dried mass-determined. The specimens are then boiled, surface-dried mass recorded, and then immersed mass recorded.	Provides useful information.	Calculation assumes void space in the specimen.
Splitting tensile strength	Determination of splitting tensile strength, which provides a measure of the tensile strength of hardened concrete specimens.	Standard prepared and cured specimens are tested in a device applying eccentric load.	Provides useful information on tensile strength.	Accuracy is required in positioning specimens and the use of aligning jig in testing machines for consistent results.
Static modulus of elasticity and Poisson's ratio	Determination of chord (Young's) modulus of elasticity and Poisson's ratio.	Standard prepared and cured specimens are loaded in compression, and longitudinal and transverse strains are measured.	Provides useful information.	A length to diameter ratio of 1.5 or greater is required for good results.

Table 6-3. NDT Methods for Determining Chemical and Physical Properties of Concrete.

Property	Purpose of test	Principle of operation	Advantages	Limitations
Break-off method	Determination of flexural strength.	Break-off of a cylindrical specimen of in-place concrete. Cores are drilled into the concrete and broken off with a commercial tester.	Safe, simple, and fast to perform, requiring only an exposed surface.	Limited by maximum aggregate size and minimum member thickness.
Nuclear gamma radiation	Determination of in-place density.	A gamma source or a detector is housed in a probe inserted into a preformed hole in the concrete to a predetermined depth. Readings are related to density by a calibration curve.	Useful information is obtained and very accurate if calibration curves are established for each project.	Results can be altered by reinforcing steel, chemical composition of concrete constituents, and sample heterogeneity.
Neutron probe	Determination of chloride content in concrete.	Detection and counting of captured activated gamma rays emitted by chloride ions.	Can detect very small concentrations of chloride content.	To avoid modeling the neutron flux distribution, an assumption must be made on the chloride ion profile in concrete. The alternative involves considerable computation.

(*Continued*)

Table 6-3. (*Continued*)

Property	Purpose of test	Principle of operation	Advantages	Limitations
Rebound hammer	Compares the quality of concrete from different areas of the specimen; estimates concrete strength on the basis of calibration curves with limited accuracy.	Spring-driven mass strikes the surface of concrete, and rebound distance is given in *R*-values; surface hardness is measured and strength estimated from calibration curves provided by the hammer manufacturer.	Equipment is lightweight, simple to operate, and inexpensive; a large amount of data can be quickly obtained; good for determining the uniformity of concrete and areas of potentially low strength.	Results affected by the condition of the concrete surface; does not give a precise prediction of strength; estimates of strength should be used with great care; frequent calibration of equipment is required.
Resonant frequency testing	Used in the laboratory to determine various fundamental modes of vibration for calculating dynamic modules; used in the field to detect voids, delaminations, foreign objects, and deterioration and to determine thickness.	A resonant frequency condition is set up between two reflecting interfaces. Energy can be introduced by hammer impact or an oscillator–amplifier–electromagnetic driver system.	Allows one to "see inside" concrete structures; can penetrate to depths of a number of feet; a newly developed transducer receiver can improve results over an accelerometer or ultrasonic transducer.	Operates in sonic range and does not have a resolution of ultrasonics, but numerous reverberations among reflecting surfaces can improve averages. Still in the developing stage.

(*Continued*)

Table 6-3. (*Continued*)

Property	Purpose of test	Principle of operation	Advantages	Limitations
Ultrasonic pulse velocity	Gives an indication of strength, uniformity, and quality of concrete; internal discontinuities can be located and their size estimated; the most widely used stress wave method for field use.	Operates on the principle that stress wave propagation velocity is affected by the quality of concrete; pulse waves are introduced in materials, and the time of arrival is measured at the receiving surface with a receiver.	Equipment is relatively inexpensive and easy to operate; accurate assessment of uniformity and quality. By correlating the compressive strength of cores and wave velocity, in situ strength can be estimated.	Good coupling between transducer and test substrate is critical; interpretation of results can be difficult; density and amount of aggregate moisture variations and presence of metal reinforcement may affect results; calibration standard required.
Ultrasonic pulse-echo	Gives an indication of the uniformity and quality of concrete. Can give the location of rebars, voids in concrete, density, and thickness.	Operates on the principle that the original direction, amplitude, and frequency content of stress waves introduced into concrete are modified by the presence of interfaces such as cracks, objects, and sections that have impedance.	Can operate where only one surface is accessible. Can operate on a dry surface or in an underwater environment. Allows one to "see" inside concrete.	Is still in developmental stage. Needs development of measurement criteria. Not presently a standard test method. Digital signal processing can improve interpretation, but data must be returned to the lab for processing at present.

(*Continued*)

Table 6-3. (*Continued*)

Property	Purpose of test	Principle of operation	Advantages	Limitations
Windsor probe	Estimates of compressive strength, uniformity, and quality of concrete.	Probes are gun-driven into concrete; depth of penetration converted to estimates of concrete strength by using calibration curves provided by the manufacturer.	Equipment is simple, durable, and requires little maintenance; useful in assessing the quality and relative strength of concrete; does relatively little damage to the specimen.	May not yield accurate estimates of concrete strengths; interpretation of results depends on correlation curves. Difficulty in removing the probes, which are often broken and damaging to concrete.

Table 6-4. NDT Methods for Determining Physical Conditions of Concrete.

Property	Purpose of test	Principle of operation	Advantages	Limitations
Absorption and permeability tests	Determination of surface absorption and permeability	Attachment of laboratory equipment to a concrete surface to create pressure difference and capillary action.	Assessment of concrete quality and curing.	Limited application and experience record.
Acoustic emission	Continuous monitoring of structure during service life to detect impending failure; monitoring the performance of the structure during proof testing.	During crack growth or plastic deformation, the rapid release of strain energy produces acoustic (sound) waves that can be detected by sensors attached to the surface of a test object.	Monitors structural response to applied load; capable of detecting the onset of failure; capable of locating the source of possible failure; equipment is portable and easy to operate. Good for load tests.	Expensive test to run; can be used only when the structure is loaded and when flaws are growing; interpretation of results requires an expert; currently largely confined to a laboratory; further work required.
Acoustic impact	Used to detect debonds, delaminations, voids, and hairline cracks.	Surface of the object is struck with an implement. The frequency through transmission time and damping characteristics of the resulting sound gives an indication of the presence of	Portable equipment; easy to perform with an auditory system; the electronic device requires more equipment.	Geometry and mass of test object influence results; poor discrimination for the auditory system; reference standards required for electronic testing.

(Continued)

Table 6-4. (*Continued*)

Property	Purpose of test	Principle of operation	Advantages	Limitations
		defects; equipment may vary from a simple hammer or drag chain to sophisticated trailer-mounted electronic equipment.		
Cover meters/ pachometers	Measure cover, size, and location of reinforcement and metal embedments in concrete or masonry.	Presence of steel in concrete or masonry affects the magnetic field of a probe; the closer probe is to steel, the greater the effect.	Portable equipment, good results if concrete is lightly reinforced. Good for location of rebars so as to avoid damage in coring.	Difficult to interpret results if the concrete is heavily reinforced or if wire mesh is present. Not reliable for cover above 4 in. (100 mm) Form ties often mistaken for anchors.
Electrical potential measurements	Indicating the condition of steel rebars in concrete or masonry; indicating the corrosion activity in concrete.	Electrical potential of concrete indicates the probability of corrosion.	Portable equipment; field measurements readily made; appears to give reliable information.	Information on the rate of corrosion not provided; access to rebars required.

(*Continued*)

Table 6-4. (*Continued*)

Property	Purpose of test	Principle of operation	Advantages	Limitations
Electrical resistance measurements	Determination of the moisture content of concrete.	Determination of the moisture content of concrete is based on the principle that the conductivity of concrete changes with changes in moisture content.	Equipment is automated and easy to use.	Equipment is very expensive and requires high-frequency specialized applications; dielectric properties also depend on salt content and temperature of the specimen, which poses problems in interpretation of results.
Fiber optics	To view portions of a structure that are inaccessible to the eye.	Fiber-optic probe consisting of flexible optical fibers; lens and the illuminating system are inserted into a crack or drilled hole in concrete; an eyepiece is used to view the interiors to look for flaws such as cracks, voids, or aggregate de-bonds; commonly used to look into areas where cores have been removed or boreholes have been drilled; examination of cavity walls and other holes in masonry.	Gives clear, high-resolution images of remote objects. A camera attachment for photos is available. A flexible hose enables multidirectional viewing.	Equipment is expensive; many boreholes are required to give adequate access. Mortar in masonry walls hinders view.

(*Continued*)

Table 6-4. (*Continued*)

Property	Purpose of test	Principle of operation	Advantages	Limitations
Gamma radiography	Estimating the location, size, and condition of rebars; voids in concrete; density and thickness.	Based on the principle that the rate of absorption of gamma rays is affected by the density and thickness of the test specimen, gamma rays are emitted from a source, penetrate the specimen, exit on the opposite side, and are recorded on the film.	Internal defects can be detected; applicable to a variety of materials; permanent record on the film; gamma-ray equipment easily portable.	Equipment is very expensive; gamma-ray sources are health and safety hazards; require access to both sides of the specimen.
Impact echo	Determination of cracking, cross-sectional properties, delamination, and honeycomb.	Monitors surface displacement resulting from interactions of transient stress waves with internal discontinuities.	Simple and efficient technique for interpretation by frequency analysis of displacement waveforms.	The wave speeds must be determined on a test object of known thickness.
Infrared thermography	Detection of internal flaws, crack growth, delamination, and internal voids.	Flaws detected by using selective infrared frequencies to detect various passive heat patterns that can be identified as belonging to certain defects. Through cracks in concrete and masonry may be detected in cold days.	Has the potential for becoming a relatively inexpensive and accurate method for detecting concrete defects; can cover large areas quickly.	Required special skills and equipment.

(*Continued*)

Table 6-4. (*Continued*)

Property	Purpose of test	Principle of operation	Advantages	Limitations
Laser interferometry	Determination of crack initiation and crack propagation and monitoring of the dependent behavior.	A hologram image is created of the object by the use of lasers. Fringe patterns represent the displacement field in the object surfaces.	Can monitor time-dependent behavior of creep cracks.	High cost. Limited experience with the procedure.
Nuclear moisture meter	Estimation of the moisture content of hardened concrete.	Moisture content in concrete is determined on the basis of the principle that materials (such as water) decrease the speed of fast neutrons in accordance with the amount of hydrogen produced in the test specimen.	Portable; moisture estimates can be made of in-place concrete.	Equipment is highly sophisticated and expensive; NRC license required to operate; moisture gradients in the specimen may give erroneous results.
Radar	Detection of substratum voids, delaminations, and embedments. Measurement of the thickness of concrete pavements.	Transmitted electromagnetic impulse signals are used for void detection.	Expedient method; can locate rebars and voids regardless of depths.	Equipment is expensive; reliability of void detection is greatly reduced if reinforcement is present; procedure still under development.

(*Continued*)

Table 5-4. (*Continued*)

Property	Purpose of test	Principle of operation	Advantages	Limitations
Radiographics X-ray and gamma ray	X-ray density and internal structure of concrete; location of reinforcement. Gamma-ray location, size, and condition of rebars; voids in concrete; density and thickness.	Based on the principle that the rate of absorption of X-rays or gamma rays is affected by the density and thickness of the test specimen, X-rays or gamma rays are emitted from a source, penetrate the specimen, exit on the opposite side, and are recorded on the film.	Internal defects can be detected; applicable to a variety of materials; permanent record on the film; gamma-ray equipment easily portable.	X-ray has limited field application because the equipment is heavy and costly; x-ray and gamma-ray sources are harmful to organic tissue; require access to both sides of the specimen.
Ultrasonic pulse velocity	Gives an indication of the strength, uniformity, and quality of concrete; internal discontinuities can be located, and their size estimated; the most widely used stress-wave method for field use.	Operates on the principle that stress wave propagation velocity is affected by the quality of concrete; pulse waves are introduced in the material, and the time of arrival is measured at the receiving surface with a receiver.	Equipment is relatively inexpensive and easy to operate; accurate assessment of uniformity and quality. By correlating the compressive strength of cores and wave velocity, in situ strength can be estimated.	Good coupling between transducer and test substrate is critical; interpretation of results can be difficult; the density and amount of aggregate, moisture variations, and the presence of metal reinforcement may affect results; calibration standards are required.

(*Continued*)

Table 6-4. (*Continued*)

Property	Purpose of test	Principle of operation	Advantages	Limitations
Ultrasonic pulse-echo	Gives an indication of the uniformity and quality of concrete. Can give the location of rebars, voids in concrete, density, and thickness.	Operates on the principle that the original direction, amplitude, and frequency content of stress waves introduced into concrete are modified by the presence of interfaces such as cracks, objects, and sections that have different acoustic impedances.	Can operate where only one surface is accessible. Can operate on a dry surface or in an underwater environment. Allows one to "see" inside concrete.	Is still in the developmental stage. There is a need for the development of measurement criteria. Not presently a standard test method. Digital signal processing can improve interpretation, but data must be returned to the lab for processing at present.
Visual examination	(a) Evaluation of the surface condition of concrete (finish, roughness, scratches, cracks, color). (b) Determining deficiencies in joints. (c) Determining the differential movements of the structure.	Visual examination with or without optical aids, measurement tools, photographic records, or other low-cost tools; differential movement determined over long periods with surveying methods.	Usually low cost; rapid evaluation of concrete.	Trained evaluation required; primary evaluation confined to the surface of the structure.

Table 6-5. Test Methods for Determining Chemical and Physical Properties of Reinforcing Steel, Pre-Tensioning Steel, and Post-Tensioning Steel on Metal Specimens.

Test method	Purpose of test	Principle of operation	Advantages	Limitations
Bend test	Determination of the ductility of the metal specimen and resistance to cracking.	Bend specimen at room temperature through 180 degrees without cracking on the outside of the bent portion.	Gives fast, accurate information. Low cost.	Samples must be removed from the structure for testing.
Breaking strength	Determination of the breaking strength of a finished strand after stress relieving and thermal treatment. Includes the determination of elongation, load at 1% extension, and yield strength.	Testing and breaking off strands in an extensometer and the production of a load elongation curve.	Provides accurate and useful information.	Samples must be removed from the structure. These are usually destructive.

(Continued)

Table 6-5. (*Continued*)

Test method	Purpose of test	Principle of operation	Advantages	Limitations
Chemical analysis	To obtain compositions of metal to identify the type of metal and to establish characteristics such as weldability, corrosion resistance, ductility, and other mechanical characteristics that can be established by having a knowledge of chemical compositions.	A sample of metal taken from the structure is chemically analyzed to establish the composition of metal.	Are the common tests to provide data on all chemical compositions of metal. Data can be used to evaluate corrosion resistance, weldability, and some mechanical properties. It can be used to establish the identity of metal. Although mainly a laboratory test, field testing can also be performed for spot testing.	It can be destructive if samples are removed indiscriminately from the structure. Experience needed to conduct the tests. Care should be taken in sampling to ensure that the specimen is not contaminated by other chemicals, water, and oil.
Deformation requirements	Determination of minimum average height and maximum and minimum spacing of deformations.	Standard measurements.	Provides useful and required information.	Usually accomplished by performing in situ tests, but samples may be taken from the structure. These may be destructive.

(*Continued*)

Table 6-5. (*Continued*)

Test method	Purpose of test	Principle of operation	Advantages	Limitations
Film thickness	Determination of the film thickness of epoxy-coated reinforcing steel bars.	Use of a nondestructive-type thickness gauge.	Provides useful and required information.	Samples must be taken from the structure. These may be destructive.
Strength of Connections (in welded or clipped deformed bar mats).	Determination of a specified static load without loosening.	Tests of connections against slipping are performed on an assembled mat by means of a spring balance.	Provides useful and required information.	Samples must be taken from the structure. These may be destructive.
Tensile test	To obtain yield strength, yield point, tensile strength, elongation, and reduction of area. Sometimes a stress–strain or load-elongation curve is also recorded during the test, which can be useful in evaluating ductility and modulus of elasticity.	A standard metal specimen is subjected to axial tension, and major mechanical properties are measured.	Gives fast and accurate results on the mechanical properties of metals.	Samples must be removed from the structure for testing. These could be destructive.

(*Continued*)

Table 6-5. *(Continued)*

Test method	Purpose of test	Principle of operation	Advantages	Limitations
Weight of coating (galvanized products)	Determination of coating weight by either stripping specimens or using a magnetic thickness gauge.	Determination of the weight of stripped specimens or measurement of the required coating thickness.	Provides required information.	Usually an in situ test. When samples are removed from the structure, they could be destructive.
Weld shear test	Determination of minimum average shear values in both transverse and longitudinal directions of a welded wire fabric.	Testing of fabric in a specially prepared jig utilizing a standard tension testing machine.	Provides required information.	Samples must be removed from the structure. These could be destructive.
Wire diameter	Determination of nominal diameter and conformance with permissible variations of a plain steel wire for concrete reinforcement.	Measurement of wire diameters.	Provides required information.	Usually an in situ test. When samples are taken from the structure, they may be destructive.

Table 6-6. NDT Methods for Determining Physical Conditions of Reinforcing Steel, Pretensioning Steel, and Posttensioning Steel.

Test method	Purpose of test	Principle of operation	Advantages	Limitations
Acoustic emission	Determination of the degree of debonding of reinforcing steel.	During fracturing of concrete around reinforcing steel bar deformations, the release of strain energy produces acoustic waves that can be detected by sensors attached to the surface of the test object.	Equipment is portable and easy to operate. It can be used for long-term monitoring. The method is nondestructive.	Expensive. Requires complex equipment. Quantitative results cannot be obtained.
Copper wire sensor	Determination of change in the strain (stress) state of prestressing steel.	The change in capacity among different prestressing tendons is recorded under changes in load. The change in strain is mutually proportional.	Easy to operate. It can be used for long-term monitoring.	Installation of copper wires is expensive and can usually be accomplished only at tendons close to the surface.
Coring stress relief	Determination of an average stress field around prestressed reinforcement.	Based on the stress relief that occurs during the removal of concrete cores and measurement of the resulting strains.	Measured stresses are fairly accurate, and the method can be used with a good degree of confidence. Quick and simple.	Expensive. Local anomalies can affect results.

(Continued)

Table 6-6. (*Continued*)

Test method	Purpose of test	Principle of operation	Advantages	Limitations
Electrochemical impedance	Monitoring corrosion of reinforcing steel.	Measurement of the impedance caused by the electrochemistry of the active corrosion cell.	Can determine the instantaneous corrosion rate in a quick and easy manner.	Approximate technique. Many measurements are required to determine the overall pattern of corrosion.
Flat jack	Direct measurement of stresses in concrete adjacent to reinforcing steel.	A slot is cut into the concrete, and a flat jack is installed into the slot, and initial displacement is restored. Strain (stress) is measured to restore the displacement field.	Accurate. Fairly simple concept to apply and use.	Usually limited to measuring uniaxial stresses, although other states of stress can be measured by using several slots to construct a Mohr circle.
Galvanic current	Study effect of chloride concentration.	The galvanic current measure of an electrochemical corrosion cell.	Accurate measurement of corrosion rates.	Many measurements are required to establish a pattern of corrosion.

(*Continued*)

Table 6-6. (*Continued*)

Test method	Purpose of test	Principle of operation	Advantages	Limitations
Half-cell potentials	Evaluation of potential measurements in reinforcing steel, which is a measure of active corrosion.	A copper–copper sulfate half-cell is attached—one leads to reinforcing steel and the other to concrete surfaces—and is moved around to record readings of potential in rebars. Other types of half cells are also available. The rate of change of a potential gradient is the indicator of corrosion.	Inexpensive; easy to use; fairly fast.	Prewetting of a concrete surface required. Connection to reinforcing steel involves minor demolition. Considerable natural variation is present in the results. The method does not indicate the corrosion rate.
Magnetic field	Determination of fractures in a reinforcing strand and deterioration of reinforcing steel.	Application of a steady-state magnetic field and use of a scanning magnetic field sensor to detect perturbances in the applied field caused by anomalies.	Good overall sensitivity to the loss of section and sensitivity to fractures.	Expensive equipment is required. The method is still in the experimental stage.
Magnetic permeability (Tensiomag)	Stresses in reinforcing steel and prestressing steel can be measured and monitored with time.	The incremental magnetic permeability is a measure of the gradient of a minor hysteresis loop. Steel tendons are magnetized, and the state of magnetization is measured.	Accurate measurement of stress in reinforcement and prestressing steel.	Measurements limited to parallel reinforcement or strand configuration and also limited by the size of the strand.

(*Continued*)

Table 6-6. *(Continued)*

Test method	Purpose of test	Principle of operation	Advantages	Limitations
Magnetoelastic force (Pressductor)	Stresses in prestressing steel can be determined.	Magnetic properties of strands change under the influence of forces. The length of wire is changed by magnetizing it, and the change in length is measured.	It can be installed at any position along the tendon.	Force measurement is dependent on the chemical composition and grain structure of the tendon.
Optical fiber sensors	Determination of the change in strain (stress) of prestressing steel.	A specially coated optical fiber is covered with a thin wire in a spiral pattern. The path of the spiral changes with axial stretching, and this presses on the optical fiber generating microbendings. These influence the permeability of light that is proportional to strain.	Easy to operate. It can be used for long-term monitoring.	Installation of a sensor into concrete is locally semidestructive, and the methods are limited to accessible tendons.
Polarization resistance	Measurement of the corrosion rate in reinforcing steel.	The corrosion current resulting from an electrochemical corrosion cell is measured.	Accurate determination of the corrosion rate of reinforcing steel.	A portable computer system required for control. Expensive.

(Continued)

Table 6-6. *(Continued)*

Test method	Purpose of test	Principle of operation	Advantages	Limitations
Radioscopy	Assessment of the quality of grouting in prestressed tendons.	Use of a miniaturized linear accelerator.	Useful method providing required information.	Expensive. Considerable equipment is required.
Suspension cable and stayed cable wire fractures by an acoustic survey	Determination of fractured wires in cables.	Acoustic sensors installed on to cables. Waves from the accelerometer monitored remotely.	The method has been field-tested on a number of bridges and provides reliable results when compared to destructive tests.	Information is limited by the number of sensors installed. Expensive. Only future breaks can be determined.
Suspension cable and stayed cable wire fractures by magnetic perturbation.	Determination of fractured wires in the cable.	Cables are scanned magnetically by a mobile unit consisting of an electromagnet and a sensor. Flaws in the wire produce magnetic perturbations from which flaw size, depth, and location can be determined.	The method has been field-tested in a number of bridges as well as in a laboratory. Can detect fractures reliably.	Limited equipment is available. Expensive. Still experimental.
X-ray permeability	Measurement of the corrosion rate of reinforcing steel.	The diffusion of ferrous ions from the surface of steel undergoing corrosion is identified by an electron probe.	The method is quantitative.	Still in the experimental stage.

Table 6-7. Evaluation of Properties of Concrete.

References: See numerical references from noted tables in Section 6.1.13.2	Table 6-2	Table 6-3	Table 6-3	Table 6-3	Table 6-2	Table 6-2	Table 6-3	Tables 6-3 and 6-4	Table 6-2	Table 6-2	Table 6-3	Table 6-3	Table 6-3
Evaluation procedure / Chemical and physical properties	Air content test	Cast-in place procedure (pullout test)	Cement content test	Electrical resistance measurements	Flexural tests	Freeze–thaw test	Gamma radiography	Nuclear moisture meter	Petrographic analysis	Test on concrete cores	Ultrasonic pulse	Windsor probe	Pulloff testing
Acidity									•	•			
Air content	•						•		•	•			
Cement content			•						•	•			
Chemical content									•	•			
Chloride content									•	•			
Compressive strength		•								•	•	•	•
Creep										•			
Density							•			•			

(Continued)

Table 6-7. *(Continued)*

References: See numerical references from noted tables in Section 6.1.13.2	Table 6-2	Table 6-3	Table 6-3	Table 6-3	Table 6-2	Table 6-2	Table 6-3	Tables 6-3 and 6-4	Table 6-2	Table 6-2	Table 6-3	Table 6-3	Table 6-3
Evaluation procedure / Chemical and physical properties	Air content test	Cast-in place procedure (pullout test)	Cement content test	Electrical resistance measurements	Flexural tests	Freeze–thaw test	Gamma radiography	Nuclear moisture meter	Petrographic analysis	Test on concrete cores	Ultrasonic pulse	Windsor probe	Pulloff testing
Elongation										•			
Modulus of Elasticity										•	•	•	
Modulus of Rupture					•								
Moisture content				•				•					
Permeability									•				
Proportions of Aggregate									•				
Pullout strength		•											

(Continued)

Table 6-7. *(Continued)*

References: See numerical references from noted tables in Section 6.1.13.2	Table 6-2	Table 6-3	Table 6-3	Table 6-3	Table 6-2	Table 6-2	Table 6-3	Tables 6-3 and 6-4	Table 6-2	Table 6-2	Table 6-3	Table 6-3	Table 6-3
Evaluation procedure / Chemical and physical properties	Air content test	Cast-in place procedure (pullout test)	Cement content test	Electrical resistance measurements	Flexural tests	Freeze–thaw test	Gamma radiography	Nuclear moisture meter	Petrographic analysis	Test on concrete cores	Ultrasonic pulse	Windsor probe	Pulloff testing
Resistance to Freezing and Thawing						•				•			
Soundness									•	•			
Splitting tensile strength										•			
Tensile strength		•			•					•			
Uniformity of mix									•				
Water–cement ratio								•	•				

Table 6-8. Evaluation of Physical Conditions of Concrete.

References: See numerical references from noted tables in Section 6.1.13.2C	Table 6-4	Table 6-5	Table 6-4	Table 6-4	Table 6-37	Table 6-4	Table 6-2	Table 6-4	Table 6-4	Table 6-3	Table 6-2	Table 6-4	Table 6-4	Table 6-3
Evaluation procedure / Chemical and physical properties	Acoustic emission/impact	Fiber optics	Gamma radiography	Infrared	Load testing	Pachometer	Petrographic analysis	Physical measurement	Radar	Schmidt rebound hammer	Tests on concrete cores	Ultrasonic pulse	Visual examination	Windsor probe
Alkali–carbonate reaction							•							
Alkali–silica reaction							•							
Bleeding channels													•	
Cement aggregate reaction							•							
Chemical deterioration													•	

(Continued)

Table 6-8. (*Continued*)

References: See numerical references from noted tables in Section 6.1.13.2C	Table 6-4	Table 6-5	Table 6-4	Table 6-4	Table 6-37	Table 6-4	Table 6-2	Table 6-4	Table 6-4	Table 6-3	Table 6-2	Table 6-4	Table 6-4	Table 6-3
Evaluation procedure / Chemical and physical properties	Acoustic emission/impact	Fiber optics	Gamma radiography	Infrared	Load testing	Pachometer	Petrographic analysis	Physical measurement	Radar	Schmidt rebound hammer	Tests on concrete cores	Ultrasonic pulse	Visual examination	Windsor probe
Chloride attack							•						•	
Contaminated aggregate							•							
Contaminated mixing water							•							
Cracking	•							•			•	•	•	
Cross-sectional properties			•					•				•		
Delamination	•		•	•					•		•	•	•	

(*Continued*)

Table 6-8. *(Continued)*

References: See numerical references from noted tables in Section 6.1.13.2C	Table 6-4	Table 6-5	Table 6-4	Table 6-4	Table 6-37	Table 6-4	Table 6-2	Table 6-4	Table 6-4	Table 6-3	Table 6-2	Table 6-4	Table 6-4	Table 6-3
Evaluation procedure / Chemical and physical properties	Acoustic emission/impact	Fiber optics	Gamma radiography	Infrared	Load testing	Pachometer	Petrographic analysis	Physical measurement	Radar	Schmidt rebound hammer	Tests on concrete cores	Ultrasonic pulse	Visual examination	Windsor probe
Deterioration			•	•			•				•	•	•	
Discoloration							•						•	
Disintegration			•	•			•				•	•	•	
Distortion													•	
Efflorescence				•			•						•	
Erosion							•						•	

(Continued)

Table 6-8. (*Continued*)

References: See numerical references from noted tables in Section 6.1.13.2C	Table 6-4	Table 6-5	Table 6-4	Table 6-4	Table 6-37	Table 6-4	Table 6-2	Table 6-4	Table 6-4	Table 6-3	Table 6-2	Table 6-4	Table 6-4	Table 6-3
Evaluation procedure / Chemical and physical properties	Acoustic emission/impact	Fiber optics	Gamma radiography	Infrared	Load testing	Pachometer	Petrographic analysis	Physical measurement	Radar	Schmidt rebound hammer	Tests on concrete cores	Ultrasonic pulse	Visual examination	Windsor probe
Freeze–thaw damage							•						•	
Honeycomb			•	•			•				•	•	•	
Leaching							•						•	
Popouts							•						•	
Scaling							•						•	
Spalling			•	•							•	•	•	
Stratification							•						•	

(*Continued*)

Table 6-8. (*Continued*)

References: See numerical references from noted tables in Section 6.1.13.2C	Table 6-4	Table 6-5	Table 6-4	Table 6-4	Table 6-37	Table 6-4	Table 6-2	Table 6-4	Table 6-4	Table 6-3	Table 6-2	Table 6-4	Table 6-4	Table 6-3
Evaluation procedure / Chemical and physical properties	Acoustic emission/impact	Fiber optics	Gamma radiography	Infrared	Load testing	Pachometer	Petrographic analysis	Physical measurement	Radar	Schmidt rebound hammer	Tests on concrete cores	Ultrasonic pulse	Visual examination	Windsor probe
Structural performance	•				•								•	
Sulfate attack							•						•	
Uniformity of Concrete			•				•			•		•	•	•
Unsound cement							•							
Unsound Concrete							•			•		•	•	•

Table 6-9. Evaluation of Properties of Reinforcing Steel, Pretensioning Steel, and Posttensioning Steel.

References: See Section 6.1.13.2 and Table 6-5	Table 6-5	Table 6-5	Table 6-5	Table 6-5
Evaluation procedure / Chemical and physical properties	Chemical analysis	Coating tests	Material specification	Tension tests
Bend test			•	
Breaking strength				•
Carbon content	•			
Chemical composition	•	•		
Coating properties				
Adhesion		•		
Continuity		•		
Thickness		•		
Deformation requirements			•	
Elongation				•
Reduction of Area				•
Strength of Connections			•	
Tensile strength				•
Weld shear strength			•	
Yield strength				•

Table 6-10. Evaluation of Physical Conditions of Reinforcing Steel, Pretensioning Steel, and Posttensioning Steel.

References	Table 6-4	Table 6-4	Table 6-4	Table 6-4	Table 6-4	Table 6-4
Evaluation procedure / Physical condition	Cover meter	Gamma radiography	Probe holes	Physical measurement	Radar	Visual examination
Anchorage	•		•	•		•
Corrosion						•
Cover	•	•	•	•	•	
Cross-sectional properties				•		
Exposure						•
Location	•	•	•	•	•	
Shape	•		•	•		

6.2 CONDITION ASSESSMENT OF METALS

The condition of a metal structure may be determined by visual examination in conjunction with computational methods and testing. Metal properties and physical conditions are discussed in Section 6.2.4. Test methods that can be used to evaluate these properties are provided in Section 6.2.13. A condition assessment of metal may require the determination of material strength. Proper assessment and subsequent evaluation (Section 6.2.2) may reveal the metal's ability to sustain loads and environmental conditions to which it is being or will be subjected.

Basic information for analyzing steel structures is found in the American Institute of Steel Construction *Steel Construction Manual* (AISC 2022 or current edition). For cold-formed steel analysis and evaluation, the reader should reference the American Iron and Steel Institute *Cold-Formed Steel Design Manual* (AISI 2017). Aluminum structure and member analysis and evaluation are found in the *Aluminum Design Manual* provided by the Aluminum Association (AA 2020). Member analysis and evaluation of stainless steel structures are found in the *AISC Steel Design Guide 27—Structural Stainless Steel* (AISC 2022). A new AISC design manual for stainless steel is underway and should be referenced when completed.

6.2.1 Reasons for Conducting a Metal Condition Assessment

The performance of metal structures is dependent on numerous factors, including strength, stiffness, and connections—all of which are subject to weakening or failure caused by deterioration or physical damage. The following are the primary concerns to be addressed in a condition assessment of metals:

1. **Moisture Concerns**: Metals (including cast iron, wrought iron, and most steels) are vulnerable to corrosion from exposure to moisture. Because of the light mass of metals relative to their strength, large deflections and instability of metal elements can result.
2. **Strength or Stiffness Questions**: The condition of the structural metal in the building should be assessed for strength and stiffness by examining the material properties and physical conditions.
3. **Connections**: Connections play a major role. Special attention should be paid to the condition assessment of connections and connectors in metal structures when performing an assessment.
4. **Fire**: Metals are vulnerable to fire if not properly protected.
5. **Deterioration Observations**: There is a need to determine the cause of deterioration, locate and quantify the extent of deterioration, and assess its impact on section properties.

6.2.2 Metal as a Material

The function of metal in a structure is to provide the strength and stiffness to carry the loads. This function requires that the chemical and physical properties of the metal remain adequate for the expected loading conditions. Metallic components and systems carry the applied loads in bending, shear, torsion, axial tension or compression, or a combination of these.

6.2.3 Terminology Related to Metal Characteristics and Properties

The following list gives common terms that are relevant to metals when conducting a structural condition assessment.

1. **Alloy Steel**: A steel, other than a stainless steel, that conforms to a specification that requires one or more of the following elements, by mass percent, to have a minimum content equal to or greater than the following: 0.30 of aluminum; 0.0008 of boron; 0.30 of chromium; 0.30 of cobalt; 0.40 of copper; 0.40 of lead; 1.65 of manganese; 0.08 of molybdenum; 0.30 of nickel; 0.06 of niobium (columbium); 0.50 of silicon; 0.05 of titanium; 0.30 of tungsten (wolfram); 0.10 of vanadium; 0.05 of zirconium; or 0.10 of any other alloying element, except sulfur, phosphorus, carbon, and nitrogen.

2. **Carbon Steel**: A steel that conforms to a specification that prescribes a maximum limit, by heat analysis in mass percent, of not more than 2.00 of carbon and 1.65 of manganese but does not prescribe a minimum limit of chromium, cobalt, molybdenum, nickel, niobium (columbium), tungsten (wolfram), vanadium, or zirconium.
3. **Cross-Sectional Properties**—Actual or nominal dimensions and other geometric properties of structural members (Section 6.2.13, Table 6-14).
4. **Direct Chemical Attack**: Deterioration of metals by attack of chemical solutions with either low or high acidity (pH).
5. **Ductility**: Mechanical property that is a measure of the ability to deform before cracking.
6. **Electrolytic or Electrochemical Corrosion**: Oxidation of metal caused by a chemical reaction between the metal and oxygen in the environment; the most common cause of deterioration of unprotected iron and steel products (rust).
7. **Elongation**: Increase in gauge length of the tension test specimen measured after a fracture as a percentage of the original length; a measure of ductility.
8. **Fatigue Cracking**: Cracks that develop in ductile metals when the material is subjected to cyclic stresses. These cracks initially propagate slowly and, if detected in time, can be treated by taking remedial action. If the cracks are allowed to propagate unrestricted, they frequently initiate brittle fracture. A procedure for identifying fatigue cracking by a liquid penetrant is illustrated in Figure 6-17. Also refer to ASTM E1417/E1417M (ASTM 2021) on the standard practice for liquid penetrant testing.
9. **Fatigue Properties**: Mechanical properties such as cyclic testing, endurance limit, *e–N* curve, hardness, and *S–N* curve that characterize the resistance of the metals to fatigue crack initiation and propagation under low-amplitude high-frequency cyclic stresses or strains.
10. **Fracture Cracking**: Brittle cracks that take place with little or no preceding plastic deformation. Low temperature (possibly as high as room temperature), stress or strain concentrations (including

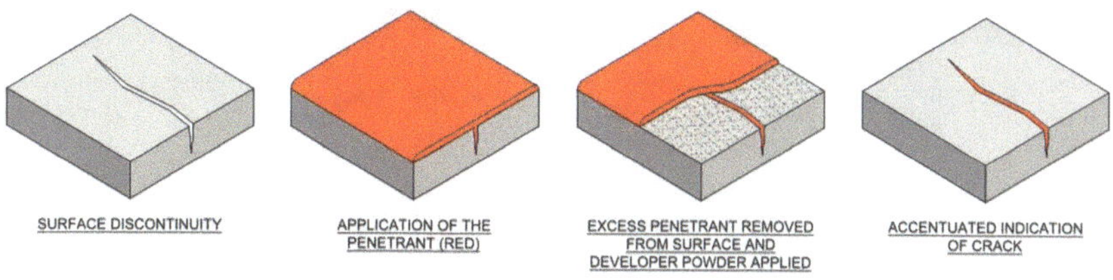

Figure 6-17. Use of a liquid penetrant to identify fatigue cracks.
Source: Courtesy of Raths, Raths & Johnson, Inc.

microcracks and fatigue cracks), and metallurgical composition are important factors influencing fracture cracking. These types of cracks (as illustrated in Figure 6-18) are often triggered by impact or sudden increases in load.

A liquid penetrant may also be used to identify fracture cracking (Figure 6-18). Internal defects and cracks can be identified by using either a normal probe or angle probe ultrasonic testing, as illustrated in Figure 6-19. Also refer to ASTM A388/A388M (ASTM 2023) on standard practice for ultrasonic testing for detection of discontinuities in steel forgings.

11. **Fracture Properties**: Mechanical properties such as Charpy number, fracture toughness, and nil ductility transition temperature that characterize the resistance of the metals to brittle fracture, especially at low temperature.
12. **Geometry of Structural Components**: Actual or nominal dimensions of structural components.

Figure 6-18. Fracture cracking.
Source: Courtesy of Raths, Raths & Johnson, Inc.

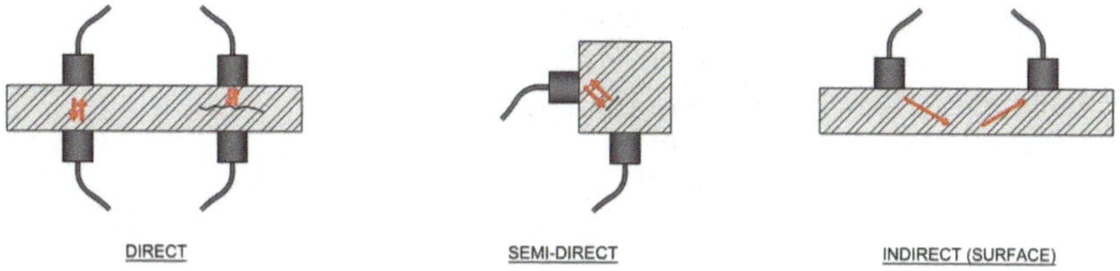

Figure 6-19. Use of ultrasonic probes.
Source: Courtesy of Raths, Raths & Johnson, Inc.

13. **Geometry of Structure**: Actual or nominal dimensions are defining the geometry of the structure.
14. **Imperfection**: A material discontinuity or irregularity that is detectable by **Inspection**.
15. **In Situ Stresses**: The actual stresses that exist in a member representing the sum of manufacturing, fabrication, and erection residual stresses and the stresses resulting from in situ loads and deformations.
16. **Inspection**: The process of measuring, examining, testing, gauging, or otherwise comparing the unit of the product with the applicable requirements.
17. **Laminar Tearing**: A planar separation that develops in flanges of rolled shapes and within thick plates near certain large welds as high-weld shrinkage develops stresses across the plate thickness.
18. **Modulus of Elasticity**: Ratio of normal stress to the corresponding strain in the linear elastic segment of a stress–strain curve.
19. **Overstressing**: Subjecting metals to stresses in excess of their allowable stress levels.
20. **Proof Load**: A specified load without a measured permanent set; applicable to some assemblies.
21. **Reduction of Area**: Decrease in the area of the fractured section of a tension test specimen as a percentage of the original area.
22. **Rotational Lateral Stiffness**: Rotational stiffness of the actual beam/panel configuration under load.
23. **Shear Strength**: Maximum shear stress that a material is capable of sustaining. Shear strength is calculated from the maximum force during a shear or torsion test and is based on the original dimensions of the cross section of the specimen.
24. **Shear Wall In-Plane Properties**: Strength and stiffness of a framed wall assembly under shear loads parallel to the wall to be used in the design of the wall.
25. **Stainless Steel**: Steel that conforms to a specification that requires, by mass percent, a minimum chromium content of 10.5 or more and a maximum carbon content of less than 1.20.

26. **Steel**: Material that conforms to a specification that requires, by mass percent, more iron than any other element and a maximum carbon content of, in general, less than 2.
27. **Strain**: Per unit change in the size or shape of a body in reference to its original size or shape. Strain is a nondimensional quantity, but it is frequently expressed in inches per inch, meters per meter, or percent.
28. **Stress**: Intensity at a point in a body of the forces or components of force that act on a given plane through the point. Stress is expressed in force per unit of area (e.g., pounds force per square inch, megapascals).
29. **Tensile Strength**: Stress calculated from the maximum load sustained by a metal specimen during a tension test divided by the original cross-sectional area of the specimen.
30. **Test Method**: Definitive procedure, preferably standardized, that produces a test result.
31. **Ultimate Loading**: Specified load at which an assembly or system fails in buckling, fracture, and so on.
32. **Yield Strength**: Stress at which a significant increase in strain occurs without the same proportional increase in stress or as defined by the appropriate standard specification for the metal. Relatively high strains (rates of loading) are used in mill tests relative to the slower rates of the essentially static loading that takes place in most structures. Because the yield point of steel increases with increasing strain rate, the values measured in mill tests are higher than those that are effective in the structure (Winter 1961).

6.2.4 Physical and Mechanical Properties of Metals

This section discusses frequently measured chemical and mechanical properties and physical conditions of existing metal structures. Properties are measured to evaluate the strength and stiffness. An examination of the physical conditions of metals can be useful in evaluating strength, stiffness, degree of corrosion, and fire protection. As a structural material, the properties and physical conditions of steel that are summarized in Tables 6-16 and 6-17 should be considered in evaluating the acceptability of existing metal members and connections.

1. **Bracing of Compression Elements and Members**: Lack of necessary bracing can result in a loss of stability, strength, and stiffness of main load–carrying elements.
2. **Carbon Content and Chemical Composition**: Indicators of properties of metals that can be used to establish corrosion resistance and mechanical properties. Both carbon content and carbon equivalent content are needed to establish weldability of the metal.
3. **Deformations**: Deformations of members in a metal structure that may be caused by fabrication or erection out-of-plumbness, lack of fit

Figure 6-20. Local buckling in a steel wide flange beam.
Source: Courtesy of Raths, Raths & Johnson, Inc.

or slip at connections, settlement or failure of supports, overstressing inadequate mechanical properties of materials used, inadequate bracing, change in temperatures, removed members or missing connectors, and torsional effects unaccounted for in the design.

4. **Strength and Stiffness**: Strength and stiffness of metals may deteriorate significantly while subjected to elevated temperatures.
5. **Global or Local Buckling**: Metal members are relatively slender and usually have thin elements, making them susceptible to global or local buckling (as illustrated in Figure 6-20).
6. **Torsional Rigidity**: Cold-formed steel members typically consist of thin elements, making them relatively weak against torsion.

6.2.5 Metal Products

Metals include all grades of hot-rolled and cold-formed structural steel, stainless steel, cast iron and wrought iron, parallel and helical steel wire structural strand, structural steel rope, steel rods, aluminum products, and other metals.

Most metal products are manufactured off-site, transported to the site, and erected into its final place in the completed structure. Some of these metal products are beams, joists, and trusses, among others. Some metal products are used in conjunction with other materials to fabricate systems off-site (wall panels, roof panels, etc.), which are then transported to the site and erected into its final place in the completed structure.

Other metals such as copper, lead-coated copper, terne-coated stainless steel, and Monel are used in the building envelope and may not be a structural element.

6.2.6 Factors That Influence Metal Behavior

The function of metal in a structure is to provide the strength and stiffness to carry the loads. This function requires that the chemical and physical properties of the metal remain adequate for the expected loading conditions. The behavior of metal, directly related to its chemical and physical properties, can be changed by such factors as exposure to water, corrosive marine and industrial atmospheres, extremely high (fire) or low temperatures, excessive loading conditions, and loss of effective compressive bracing (usually associated with physical damage or alterations of the bracing components and systems). These causes are interrelated in the sense that more than one mechanism may be prevalent in the deterioration process at the same time.

6.2.7 Mechanisms of Metal Deterioration

Deterioration of metals can be caused by one factor or a combination of many factors. Ultimately, metal deterioration from these factors can typically be categorized as corrosion, fatigue, stress corrosion, or damage from extreme events. Fracture, or a break in the mechanical continuity, is the end result of deterioration.

Corrosion: Atmospheric corrosion (to distinguish the mechanism from crevice corrosion, pitting corrosion, stress corrosion, etc.) is, in general, an electrochemical process (Mattsson 1982). This process takes place in small local galvanic cells, which are similar to small batteries with anodes and cathodes acting as electrodes. These cells operate only in the presence of an electrolyte, which means that corrosion occurs only when the surface is wet. This moisture film will contain various materials deposited from the atmosphere as well as those originating from the corroding metal. Oxygen will be readily absorbed from the air so that at least the outer region of the thin water film on the metal surface may be considered saturated with oxygen. Sulfur oxides, nitrogen oxides, and chlorides will also be absorbed from the atmosphere. The concentrations of these various materials absorbed from the atmosphere in the moisture film serve as the electrolyte. The difference in potential results from the fact that different metal surfaces possess different solutions or electrode potentials. This difference in potential results in flow of electric current and causes the local attack to begin. The corrosion rate (usually considered as the rate of metal loss or corrosion cement density) depends on many factors, including the amount and composition of the electrolyte on the surface, the relative area of the anodes and cathodes, and the potential difference between them.

Steels with improved resistance to corrosion are called *weathering steels* and include stainless steels and high-alloy steels. Rust developing on

Figure 6-21. Steel pile buckling caused by corrosion.
Source: Courtesy of Exponent.

weathering steels is more protective than rust on unalloyed or carbon steel. The penetration of rust on weathering steel approaches a nearly constant value after some years. This layer of rust then offers a protective coating known as a *patina*. However, when the atmosphere is highly polluted with sulfur dioxide and under climatic conditions with prolonged wet periods, weathering steel behaves much like unalloyed carbon steels. The same is true for weathering steels in marine atmospheres, especially in sheltered positions, in tunnels, on soil, on concrete, and in immersion situations.

Atmospheric corrosion is often the initial stage of deterioration of metal, weakening the member and making it susceptible to mechanical failure or leakage, fatigue, or brittle cracking and stress corrosion. An example of steel pile buckling resulting from corrosion is illustrated in Figure 6-21.

Fatigue: ASTM International defines fatigue as "the process of progressive localized permanent structural change occurring in a material subjected to conditions that produce fluctuating stresses and strains at some point or points, and that may culminate in cracks or complete fracture after a sufficient number of fluctuations." Other significant ASTM definitions relating to fatigue are

- *Fatigue Life, N*: Number of cycles of stress or strain of a specified character that a given specimen sustains before failure of a specified nature occurs.

- *Fatigue Limit*, S_t: Limiting value of the median fatigue strength as N becomes very large. Also, the stress range below which fatigue is not expected to occur regardless of the number of cycles.
- *S–N Diagram*: Plot of stress range against the number of cycles to failure.

Some of the factors known to affect the fatigue strength (Munse 1964) include

- Stress concentrations;
- Number of cycles or repetitions of load;
- Magnitude, nature, and range of applied stress;
- Material properties;
- Size effects;
- Surface finish;
- Rate of load application;
- Residual stresses;
- Rest periods; and
- Temperature.

Present design procedures treat all these factors adequately so that theoretically, there should be no fatigue problem. This is not true for older, existing structures that may have been designed and constructed without adequate consideration having been given to fatigue. Even with new buildings, details may be used that have not been properly evaluated for fatigue life, and fatigue limits and flaws may exist in materials, or be introduced during the fabrication and construction processes, that create stress concentrations above values anticipated.

The most important parameters affecting fatigue are the number of load repetitions and magnitude and range of stress resulting from these repetitions. For specific materials and design details, the fatigue life, *N*, and the fatigue limit, S_t are determined in the laboratory to develop an *S–N* diagram for the specific material and detail. Based on the results of these tests, design specifications limit allowable stresses or load factors for members especially susceptible to fatigue, termed fracture critical members, and for members with specific fatigue-susceptible details. Accordingly, most fatigue failures develop because of stress concentrations caused by the nature of specific details or by defects or residual stresses in the material or component that result from the manufacturing, fabrication, and construction processes.

Fatigue cracking and subsequent fracture occurs in the tension zones of structural members and may be induced by displacement (the geometry of the detail allows excessive distortion with resulting stress concentration), by initial defects or geometry, or by aeroelastic action. Atmospheric corrosion can play a part by reducing the cross sections of the resisting elements, thereby increasing the magnitude of stress over a given range. Unanticipated levels of stress over a stress range in a tencted cycles of load may result in fatigue cracking and fracture.

Stress Corrosion: Fracture caused by stress corrosion is not a common problem in buildings and is usually associated with specific connection details such as eye bars and pin connections in bridges. A rare example is the Berlin Congress Hall Roof (Isecke 1982). Like fatigue cracking and fracture, stress corrosion is usually initiated by flaws or defects in detail, which generate high concentrations of stress along with atmospheric corrosion, which reduces member sections and thereby increases stresses. With these conditions, alloy steel tensile members in the right environment can fracture from stress corrosion.

Damage from Accidents: Although not an obvious cause of structural steel deterioration, damage from accidents or extreme events is not uncommon. This usually results from structural members being struck by vehicles in building garages and loading areas or by ships berthing at wharves supporting buildings. Structural members can also be damaged by fire, impacts, storms, and other extreme events. The fire protection system for steel is usually affected by physical damage or alterations and should be restored accordingly to maintain the required fire rating of the steel.

6.2.8 Considerations with Metals Used in Historic Construction

Metals commonly found in existing buildings for structural purposes are cast iron, wrought iron, steel (ferrous or stainless), and aluminum alloy. Cast iron and wrought iron are very rarely used today as construction metals; however, cast and wrought iron members can be found in many buildings built in the nineteenth and early twentieth centuries. Cast-iron columns began to replace timber columns in mill buildings in the early 1800s and became common in other building structures by midcentury. Their use continued until the early part of the twentieth century. Cast-iron beams were, in general, asymmetric, with large bottom flanges reflecting the low tensile strength and flexural reliability of the material. Cast-iron arches and arch assemblies of cast-iron panels used tension rods of wrought iron. Building fronts of cast iron consisted of columns, panels, and ornamental elements such as cornices, column capitals, dormers, windows, and major enclosing plates on mansard roofs. Cast and wrought iron materials must be carefully analyzed because of known tendencies to behave poorly under high tensile and shear loadings.

As rolling techniques were developed for wrought iron, various shapes and plates were combined into built-up sections for columns, beams, and trusses. That progressed into full-size shapes with good flexural reliability. The transition from wrought iron to steel occurred over some years and had been mostly complete by the mid-1890s.

Cold-formed steel members have been used in building construction increasingly since the end of World War II. Today, cold-formed steel bearing walls, floor systems, and truss assemblies in multistory buildings are accepted as engineered systems. Cold-formed shapes are, in general, formed from galvanized steel sheets. Exposed steel roof decks are usually protected by electroplated metals or an applied coating.

Weathering steel (ASTM A588) is mostly used for bridge structures, although some early applications were used for exposed building structures. The US Steel Tower in Pittsburgh, Pennsylvania, completed in 1971, is an example of a building using exposed weathering steel.

Stainless steel has been used on buildings for many years as ornamentation and façade elements exposed to the weather. More recently, austenitic (300 series) stainless steels have been utilized structurally where resistance to corrosion is of concern.

Aluminum alloys (usually 6061 or 6063) have been employed structurally in buildings for nearly 100 years. The advantages are strength and light weight and the disadvantages are increased deflections and corrosion.

For about 150 years, corrosion resistance for ferrous metals has been provided by coating with molten zinc. A light coating is applied to steel coils by electroplating. Heavier coatings are achieved by hot dipping in a kettle of molten zinc.

6.2.9 Connections

Connections include the component metal shapes, wire ropes, wire strands, and rods used to attach one structural element to another. Connectors include welds, rivets, bolts and rods, studs, and clamps. They are required to transmit forces between the elements that are being connected. As structural materials, the properties and physical conditions of connectors that are summarized in Tables 6-18 and 6-19 should be considered in evaluating the acceptability of existing connectors.

6.2.10 Terminology Related to Connector Characteristics and Properties

The following list gives common terms that are relevant to connectors when conducting a structural condition assessment:

Connection: Combination of structural elements and joints used to transmit forces between two or more members.

Fastener: Mechanical device, weld, crimp, or other similar method or device used to connect two or more elements.

Hardness: Resistance of a material to deformation, particularly permanent deformation, indentation, or scratching.

6.2.11 Physical and Mechanical Properties of Connectors

The properties listed in Sections 6.2.3 and 6.2.4 also apply for connections that consist of metal components.

Breaking Strength: Measure of the strength of wire strand and wire rope fabricated from individual wires of varying tensile strengths. It is the equivalent tensile strength of the wire strand or rope.

Chemical Composition: Indicators of the properties of bolts (including washers and nuts), rivets, and studs that can be used to establish corrosion resistance and mechanical properties.

Coating Mass: Specified requirement for wire strand and rope, bolts, and other metal connectors to ensure an appropriate level of protection against corrosion.

Condition: Physical state as determined by observation and with the use of simple physical assistance such as cleaning, scraping, and sounding; can be used to establish the existence of many of the physical conditions included herein.

Corrosion: Same as Electrolytic or Electrochemical Corrosion in Metals Characteristics and Properties (Section 6.2.3).

Cracks: Discontinuities in the weld filler material or adjacent base metal in welded joints. The extent of cracks in welded joints may be ascertained by acid etching, magnetic particle inspection (ASTM E709-21 shown in Figure 6-22), or other equally reliable means.

Cross-Sectional Properties: Actual dimensions of connectors.

Deformation: Elongation, bending, or twisting of bolts, rods, rivets, and studs over and above the elastic limit.

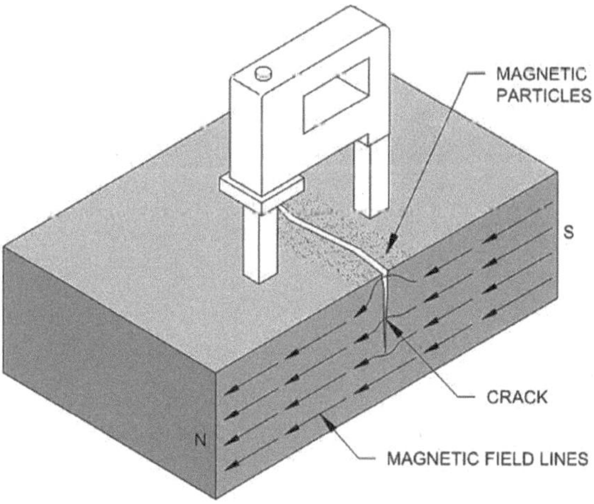

Figure 6-22. Magnetic particle inspection.
Source: Courtesy of Raths, Raths & Johnson, Inc.

Dimensions: Actual diameter and length of bolts, rods, rivets, and studs.

Discontinuities: Either internal or external localized areas of interrupted filler material in welded joints that result from cracking, entrapped slag, inclusions, deoxidation products, gas pockets, or blowholes.

Elongation: Requirement for studs, wire strands, and wire ropes measuring the increase in gauge length in a tension test specimen measured after a fracture as a percentage of the original length; a measure of ductility.

Hardness: For rivets and threaded fasteners, sufficient hardness ensures that the surfaces of the fasteners have the capacity to transmit load by bearing; a specified requirement for rivets, nuts, and washers, and sometimes for bolts and rods. Hardness is measured by a standard indicator using standard procedures, as shown in Figure 6-23.

Length: Actual physical dimension of a length of weld or a length of wire strand or wire rope.

Location: Physical position of a weld relative to the parts it connects.

Nonbinding Condition: Bolts or rods that are not fully restrained from movement by tight-fitting nuts.

Porosity: Existence of gas pockets or blow holes in the weld filler material.

Profile: Plot of the actual cross section through a weld connector. The profile can be flat, convex, or concave.

Proof Load: Specified load without a measured permanent set; applicable to bolts and rods.

Reduction in Area: Decrease in the area of the fractured section of a tensile test specimen as a percentage of the original area; a

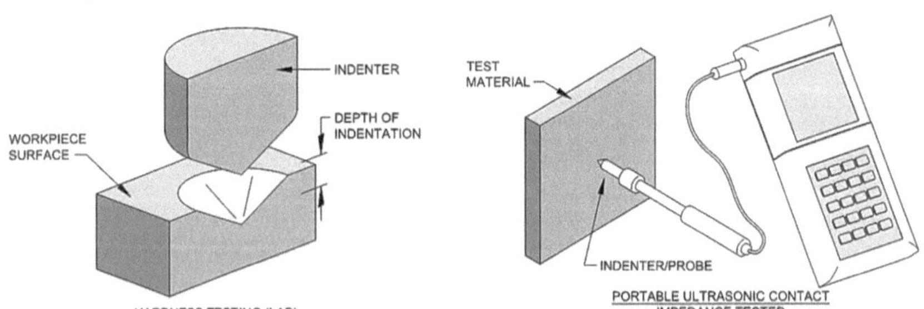

Figure 6-23. Hardness test.
Source: Courtesy of Raths, Raths & Johnson, Inc.

specified minimum requirement for studs to ensure ductility, also commonly known as *necking*.

Size: Actual dimensions of a weld used to determine its capacity; size includes the angle of the bevel, depth of penetration, and root opening for butt welds. Fillet welds are sized by the length of the leg of the weld.

Slag Deposit: Discontinuity in a welded joint resulting from concentrations of impurities in the base metal as it is fused and from the deposit of electrode-coating material in the weld.

Slotted Holes for Movement: Slotted or oversized holes that permit connector movement primarily for thermal changes in the geometry of the structure. They are also used under certain conditions to control the buildup of stresses in a connection.

Smoothness: Lack of abrupt and singular discontinuities in the weld profile.

Soundness: Physical condition of rivet connectors that can be determined by striking the rivet with a hammer and measuring by the pitch of the resulting impact.

Stress at 0.7% Extension: Mechanical requirement for wire strand and wire rope to establish the limiting load without fracturing individual wires.

Tensile Strength: Stress calculated from the maximum load sustained by a metal connector during a tension test divided by the original cross-sectional area of the specimen; a required physical property for bolts, rods, studs, wire strand, and wire rope.

Tensile Strength of Filler Material: Same as Tensile Strength for Metal Characteristics and Properties (Section 6.2.3) and a requirement of the physical property of electrodes used in welding.

Tightness: The physical condition of bolts, rods, rivets, wire strands, and wire ropes indicates that the connector fits snugly and, in general, under some positive compressive force.

Uniformity: A measure of the consistency of the weld profile, weld size, weld dimensions, weld cross-sectional properties, and conditions.

Yield Strength of Filler Material: Same as Yield Strength for Properties of Metals (Section 6.2.3) and a required physical property of electrodes used in welding.

6.2.12 Mechanisms of Connector Deterioration

Connectors are typically metal and are susceptible to the same methods of deterioration as metals, as discussed in Section 6.2.7.

6.2.13 Tabulation of Test Methods

6.2.13.1 Test Methods (Preliminary Assessment Chart). The relative value to the assessment professional of visual, nondestructive, and destructive field testing is depicted in Table 6-11 This chart is a quick reference for the applicability of certain test methods to determine metal and metal connection conditions and properties. Subsequent tables list the capabilities and limitations of various test methods in greater detail, along with a list of references that have additional information. Condition assessment of structures is dynamic, and the investigator should recognize that some technologies listed may become unavailable as devices are removed from the market or replaced by newer technology. The most common test procedures such as requirements, advantages, and limitations are listed in Tables 6-12 to 6-15 It may be necessary to use a combination of methods for satisfactorily predicting the properties of metals (Section 6.2.1.3).

Sample Size—Several ASTM standards exist to guide an investigator in determining sample size if a statistically significant sample is desired. Often a statistically significant sample is not available or economically viable. In this case, it is desired to select a sample without bias. This sample could also be a random sample, but not necessarily, as purely random samples in a failure investigation are often not practical given the constraints of the existing structure. As sample size and distribution are determined, care should be exercised to obtain samples from damaged and undamaged and questionable and unquestionable portions of the structure if applicable.

ASTM E122, *Standard Practice for Calculating Sample Size to Estimate, with a Specified Tolerable Error, the Average for a Characteristic of a Lot or Process* (ASTM 2017), can be followed to determine the number of samples required for a desired level of accuracy. The number of samples depends on

- Error E that can be tolerated in the estimate of the measured value;
- Preferred probability, or confidence level, that the measured value is within the limit of E; and
- Estimate of the variation, σ, of the measured value from one unit to another within the population.

Assuming that the measured result is normally distributed and that the population is large, the following equation, based on an assumed 95% confidence interval, can be used to calculate the number of samples, n, for testing:

$$n = \left(\frac{1.96\sigma}{E}\right)^2 \tag{6-2}$$

Table 6-11. Relative Assessment Value for Metal and Connection Conditions and Properties.

Factor	In situ visual inspection	Bolt tightness	Ultrasonic pulse velocity	Magnetic particle	Eddy current	Acoustical emission	Liquid penetrant	Sampling and testing	Destructive openings	Radiography
Material properties metals								Yes		
Locate deterioration/defect	Limited		Yes						Yes	Yes
Quantify deterioration/defect	Limited	Yes	Yes						Limited	Yes
Locate stress concentrations						Yes				
Assess strength								Yes		
Determine modulus of elasticity								Yes		
Identify hidden defects					Limited				Limited	
Locate/assess connections	Yes/limited			Yes			Yes	Yes	Limited	Yes

where

n = Number of samples,
σ = Standard deviation of the measured results, and
E = Allowable error in the measured results.

The factor 1.96, based on a desired 95% confidence level, gives a 0.050 probability that the actual value is not within plus or minus the allowable error, E, of the measured property.

Deterministic sampling describes a situation where a set number of samples are available, or sampling is constrained by any of the factors discussed herein, and only a set number of samples can be taken (access, effects on the element under consideration, budget, etc.). This sampling protocol requires engineering judgment to ensure that a representative estimate of the characteristic under consideration is obtained.

6.2.13.2 Test Methods (Comprehensive Assessment Charts). Testing of specimens from the members of a structure to determine chemical and physical properties can provide useful quantitative data. Although removal of specimens may be destructive, careful selection and handling can ensure that the usefulness of the member is not impaired, and the damaged areas can be satisfactorily repaired. Tests on specimens provide specific, quantitative information without relying on interpretation or analytical results. Consideration should be given to physical testing of specimens over nondestructive testing methods to determine similar information. It is also advantageous to use the results of physical tests in specimens from members as base data to corroborate nondestructive testing results. Most nondestructive methods require highly qualified and skilled personnel, specialized equipment, and considerable analysis and interpretation of data to provide qualitative and quantitative information.

Available test methods are cited in Tables 6-11 through 6-19. The tables are not intended to provide the reader with information and details on how to perform and interpret tests but rather to provide information on test methods that are available, what they can accomplish, advantages and limitations of the tests, and a brief statement about qualifications of test personnel. This information is not exhaustive and is constantly subject to change. Advantages, limitations, and costs can change with time. Accordingly, the references are the most important part of these tables, and care has been taken to provide the most current and in-depth information. Perusal of any specific reference will usually lead to an extensive bibliography on the subject. The supplemental references in Appendixes B2 and B4 contain information that can be perused for additional knowledge.

Table 6-12. Test Methods for Determining Chemical and Physical Properties of Steel on Metal Specimens.

Property	Purpose of test	Principle of operation	Advantages	Limitations
Metallic chemical analysis	Determination of metal compositions to identify the type of metal and establish characteristics such as weldability, corrosion resistance, ductility, and other mechanical characteristics.	A sample of metal is chemically analyzed to establish the composition of metal.	Common tests to provide data that may be used to evaluate corrosion resistance, weldability, and some mechanical properties. Although mainly a laboratory test, field testing can also be performed for spot testing.	Can be destructive if samples are removed indiscriminately from the structure. Experience needed to conduct the tests.

(*Continued*)

Table 6-12. (*Continued*)

Property	Purpose of test	Principle of operation	Advantages	Limitations
Compression	Determination of modulus of elasticity, yield strength, yield point, and compressive strength.	A test specimen is taken from the structure and prepared. It is subjected to increasing axial compressive load, with both load and strain monitored.	A common test used to determine mechanical properties; tension test is more common.	Buckling of the specimen must be controlled, and barreling of the end section may result in nonuniform transverse deformation.
Fastener mechanical properties	Determination of the mechanical properties of threaded fasteners, washers, and rivets, including product hardness, proof load, and axial tension.	Samples of fasteners are required for testing.	High degree of accuracy because fasteners are usually available. Tests can be considered nondestructive.	Expensive. Information limited by the number of samples tested.

(*Continued*)

Table 6-12. (*Continued*)

Property	Purpose of test	Principle of operation	Advantages	Limitations
Fatigue properties	Determination of the fatigue characteristics of metals. Common fatigue properties in high-cycle fatigue are *S–N* curve, *e-N* curve, endurance limit, and constant life diagram. The major properties in low-cycle fatigue are hysteresis loops, cyclic strain hardening exponent, and the cyclic strength coefficient.	Several tests are conducted to establish the fatigue properties of metals. In these tests, standard specimens are subjected to axial, torsion, or bending cyclic strain or stress. In high-cycle low-amplitude fatigue, the amplitude of stress or strain is lower than yield point, and a large number of cycles are required to cause fatigue damage. In low-cycle high-amplitude fatigue, the amplitude of strain is higher than yield strain. In this case, a small number of cycles can cause fracture because of low-cycle fatigue.	Provides useful information on the fatigue resistance of metals.	High costs; difficulty in obtaining nondestructive samples from the structure; results sometimes cannot be directly applied to the field conditions unless service conditions are the same as test conditions.

(*Continued*)

Table 6-12. *(Continued)*

Property	Purpose of test	Principle of operation	Advantages	Limitations
Fracture	Determination of fracture properties such as plane strain fracture, toughness, R-curve, J-integral, nil-ductility transition temperature.	Standard notched specimens are subjected to monotonic or impact loading, causing fracture of the specimen through the notch. Using load deflection, load-crack tip opening displacement, and other recordings during the test, the fracture properties of metal are established. The properties can be used to estimate the resistance of metal to fracture.	Moderate cost; gives quantitative measures of fracture toughness to be used in analysis. Qualified laboratories can perform the test.	Difficulties in obtaining nondestructive samples from the structure. Careful specimen preparation is needed. Test results are sensitive to operation and are affected by residual stresses.

(Continued)

Table 6-12. (*Continued*)

Property	Purpose of test	Principle of operation	Advantages	Limitations
Hardness	Determination of hardness using a hard steel ball or diamond-shaped penetrator is forced into the material using standard procedures. By measuring the geometry of the indentation mark left on the material, hardness number is established. Several hardness numbers are defined and used in industry. Brinell and Rockwell hardness numbers are used frequently to define the hardness of metals.	Evaluate resistance to permanent or plastic deformations. To obtain a quick but approximate value of tensile strength. Also used to evaluate the quality level and uniformity of heat treatment or cold working.	Low cost; simple; portable units available; can be adapted to many comparative examinations.	Surface preparation may be necessary. Minimum thickness and width limitations exist. Assessment will be mainly qualitative.

(*Continued*)

Table 6-12. (*Continued*)

Property	Purpose of test	Principle of operation	Advantages	Limitations
Impact	Charpy and Izod tests are used to determine the susceptibility of material to brittle fracture. Provides a comparative and relative measure of fracture toughness of material by measuring impact energy, lateral expansion, and fracture appearance. The drop weight method is used to establish nil-ductility transition temperature.	A standard notched sample of metal is fractured under the impact energy of a dropping mass. The energy to cause fracture is a comparative measure of ductility and fracture toughness. Charpy V-notch beam and Izod notched cantilever tests are two common impact tests. Another common impact test is the drop weight test that is employed to establish nil-ductility transition temperature.	Relatively simple and inexpensive method to assess fracture toughness and resistance to shock loads. A fractured surface gives an indication of the ductile or brittle nature of fracture.	Difficulties in obtaining nondestructive samples from the structure for testing. Test results are sensitive to composition and heat treatment of material and other operations and are affected by residual stresses.
Modulus test (usually part of tension test)	Determination of Young's modulus, tangent modulus, and chord modulus of metallic materials.	The test specimen is loaded uniaxially and load and strain measured. The appropriate slope is calculated from the resultant stress–strain curve.	Provides fast and accurate results.	Samples must be removed from the structure.

(*Continued*)

Table 6-12. (*Continued*)

Property	Purpose of test	Principle of operation	Advantages	Limitations
Tension	Determination of yield strength, yield point, tensile strength, elongation, and reduction of area. Sometimes a stress–strain or load elongation curve is also recorded during the test, which can be useful in evaluating ductility and modulus of elasticity.	A standard metal specimen is subjected to axial tension, and major mechanical properties are measured.	Provides fast and accurate results on the mechanical properties of metals.	Samples must be removed from the structure for testing.

Table 6-13. NDT Methods for Determining Chemical and Physical Properties of Steel.

Property	Purpose of test	Principle of operation	Advantages	Limitations
Hardness	Determination of hardness by using a hardness tester (sometimes portable); a hard steel ball or diamond-shaped penetrator is forced into the material using standard procedures. By measuring the geometry of the indentation mark left on the material, the hardness number is established. Several hardness numbers are defined and used in industry. Brinell and Rockwell hardness numbers are used frequently to define the hardness of metals.	Evaluate resistance to permanent or plastic deformations. To obtain a quick but approximate value of tensile strength. Also used to evaluate the quality level and uniformity of heat treatment or cold working.	Low cost; simple; portable units available; can be adapted to many comparative examinations. A permanent record is obtained.	Surface preparation may be necessary. Minimum thickness and width limitations exist. Assessment will be mainly qualitative. Access to the structure and means of the holding tester may be limited.

Table 6-14. NDT Methods for Determining Physical Conditions of Steel.

Test method	Purpose of test	Principle of operation	Advantages	Limitations
Acoustic emission	Determination of the locations of high-stress concentration or plastic yielding while the structure is stressed. Can be used in proof loading to monitor and detect areas of high-stress intensity.	The structure to be investigated is subjected to a mechanical or external stimulus that produces the release of strain energy. The rate of release of strain energy will be relatively high at crack tips and at local yield points. The release of strain energy generates high-frequency acoustic stress waves that can be detected by acoustic sensors attached to the structure.	Nondestructive only if the required loading does not cause damage. Can be used to detect stress raisers and local yield areas. Can be used in the long-term monitoring of the structure under loading.	Expensive; requires complex equipment. It can only detect active flaws that release strain energy under applied stress. Cannot detect static flaws that are not experiencing yielding or fracture slips. Is irreversible, and by unloading and reloading the structure to the same stress level, no emission will be produced.

(*Continued*)

Table 6-14. (*Continued*)

Test method	Purpose of test	Principle of operation	Advantages	Limitations
Bolt tightness	Determination of the effectiveness of friction of bolts.	Torque wrenches (calibrated in the laboratory) are used to measure torque, which is converted to bolt tension by using tables.	Nondestructive, easy, and inexpensive.	Not too reliable. Difficult to calibrate wrenches and to relate to tension.
Electromagnetic (Eddy current)	A means of sorting metals to detect surface discontinuities, dimensions, cracks, seams, variations in alloy composition, or heat treatment. Used to evaluate the condition of wire, tubing, for thickness gauging; for electrically conductive or magnetically permeable metals.	Is based on the change of impedance of a probe coil placed in contact with metal. The impedance of the coil probe is dependent on the material properties of the metal specimen and its composition.	Nondestructive, moderate cost; readily automated; portable; permanent records available if needed; can be adapted to many comparative analyses.	Shallow penetration; reference standards are often necessary; no absolute measurements; only qualitative measurements. Laminar cracks not open to the surface and crack planes perpendicular to the axis of an eddy current cylinder may escape detection.

(*Continued*)

Table 6-14. (*Continued*)

Test method	Purpose of test	Principle of operation	Advantages	Limitations
Liquid penetrant	To detect discontinuities open to a surface. Used to detect small cracks, lamination, weld slag, incomplete fusion in welds that are open to the surface, poor bond, gauges, porosity, laps, and fabricating discontinuities.	A liquid penetrant is applied to the surface. The liquid will penetrate discontinuities open to surface. The surface is cleaned, and a developer is applied. Because of capillary action, the liquid rises to the surface and marks the flaw or crack on the developer. Two types of liquid penetrants are used: a dye penetrant that can be detected visually and a fluorescent penetrant that can be detected in darkness or when exposed to ultraviolet light.	Nondestructive; simple; inexpensive; portable; allows an inspection of complex shapes in one single operation. Applicable to magnetic or nonmagnetic material. Can be used to verify the results of magnetic particle testing. An indication of the depth of a crack can be obtained by examining the width of the bleedout.	Only defects open to the surface can be detected. Irrelevant indications can occur because of surface conditions. The temperature of the sample, penetrant drain time, emulsifier soak and drain time, drying temperature, and developing powder dwell time must be controlled carefully.

(*Continued*)

Table 6-14. (*Continued*)

Test method	Purpose of test	Principle of operation	Advantages	Limitations
Magnetic particle	To locate surface and slightly below surface discontinuities such as cracks, inclusions, lamination, porosities, voids, and laps.	The area to be inspected is magnetized. Then the area is covered by a fine powder of magnetic material. A visible pattern of discontinuity will develop because of stronger attraction at those points.	Nondestructive; simpler than radiographic inspection; relatively inexpensive. Senses surface flaws and subsurface flaws down to about 0.25 in. (6 mm) below the surface. Cracks filled with foreign objects can be detected.	Applicable only to magnetic metals. It is messy. Careful surface preparation is required. Irrelevant indications can occur. Depends on the operator's ability to interpret the results. Defects and discontinuities parallel to a magnetic field may not give pattern. Applications along two perpendicular axes may be necessary. Demagnetization after the test may be needed.

(*Continued*)

Table 6-14. (*Continued*)

Test method	Purpose of test	Principle of operation	Advantages	Limitations
Radiography	To measure microscopic voids, porosity, inclusions, and cracks. In welding, it can be used to detect blow holes, nonmetallic inclusions, incomplete penetration, and undercutting, in addition to detecting the aforementioned defects.	Is based on the principle of radiography. The x-ray or gamma-ray introduced to sample defects at voids or low-density areas. As a result, dark spots will appear at these areas on radiograph films.	Nondestructive; detects both internal and external flaws; portable; provides a permanent record on an x-ray film.	High cost; heavy; health hazard; not sensitive to defects of less than 2% of the total thickness of the specimen. Complex shapes are difficult to analyze. Very small cracks or cracks not essentially parallel to the radiation beam, such as lamination cracks, cannot be detected.

(*Continued*)

Table 6-14. (*Continued*)

Test method	Purpose of test	Principle of operation	Advantages	Limitations
Residual stress by hole drilling	To measure principal stresses on the surface of the material.	A hole is drilled in the area where residual stresses are to be measured. The stresses in the vicinity of the hole are relaxed. By attaching a three-element rosette strain gauge, the value of relaxed stresses is measured. The relieved surface strain is related to principal stresses that were present before drilling the hole.	Semidestructive with only damage caused by a small hole and cemented strain gauges. A relatively simple method to measure stresses in existing structures.	Applicable only to easily accessible surfaces where strain gauges can be properly mounted; interpretation of residual versus load-induced stresses is difficult.

(*Continued*)

Table 6-14. (*Continued*)

Test method	Purpose of test	Principle of operation	Advantages	Limitations
Ultrasonic	To detect cracks, voids, porosities, laps, poor brazing and bounding, and segregated inclusions, including flaws too small to be detected by other methods.	Ultrasonic waves are applied to the surface of metal. The waves will propagate in the metal and will be reflected by any discontinuity or change in density. The reflected echo is analyzed to locate internal flaws and cracks.	Nondestructive; portable; locates very small surface and subsurface discontinuities; instant results; accurate measure of thickness. Because of good penetration, it can be used to inspect thick objects. Extremely small flaws can be detected.	Sensitivity is reduced by a rough surface; odd shape samples are hard to analyze; requires a skilled operator. Depends on the orientation of defects and the operator's ability to interpret results. Must couple the transducer to the surface of the specimen carefully. Permanent record is not obtained. Referenced standards may be needed for calibration of equipment and for interpretation of results.

(*Continued*)

Table 6-14. (*Continued*)

Test method	Purpose of test	Principle of operation	Advantages	Limitations
Visual	To evaluate the surface condition. To check dimensional accuracy. To detect corrosion, cracks, voids, fabricating discontinuities, weld undercuts and overlaps pits, and other surface irregularities. To evaluate the condition of connections and connectors, missing or loose bolts, slag inclusions, and unwelded and poor weld appearances.	A visual examination of the surface condition of metals can be performed by the naked eye or by using special devices such as magnifiers, a borescope, fiber optics, a flashlight, a weld gauge, and a panoramic camera. Magnifiers can be used to detect small flaws and cracks.	Nondestructive; simple; inexpensive; portable; a large amount of data on the surface condition and the overall appearance of members can be obtained. Can be used to locate problem areas. Hidden surfaces can be inspected using special devices. Should be used in any evaluation even if other more powerful methods of testing are used.	Basically, qualitative information is obtained. Little quantitative measurements can be made. Only surface conditions can be evaluated. Some permanent records can be obtained by photography and recording the condition.

Table 6-15. Test Methods for Physical Properties of Steel Components, Assemblies, and Systems.

Test method	Purpose of test	Principle of operation	Advantages	Limitations
Full-scale column member tests	A full-scale column is tested to obtain the ultimate load capacity and the failure mode.	The specimen is loaded until failure, and the failure mode is recorded. Deflections and rotations can be recorded at loading increments.	The test results can be used to compare experimental to theoretical methods.	Few testing agencies have the testing equipment available to test full-scale sections.
Mechanical Properties of full sections	Short specimens are tested, and mechanical properties are calculated from the test results.	Specimens are tested in tension and/or compression to obtain the mechanical properties of the section. The tests vary on the basis of the type of field loading conditions.	The test can be used for acceptance purposes on the basis of a minimum number of tests per ton of sections made.	The manufacturer must reliably indicate that the tests represent the yield point of the sections when subjected to the kind of stress under field conditions.
Mechanically fastened connection tests	The test determines the strength and deformations of mechanically fastened connections subjected to shear and tension loads.	The actual field connection configuration is replicated and tested for the appropriate loads.	The test results will provide capacities of the connections.	The test is not intended to represent all connection configurations. The geometry of fasteners may limit the connections' capacity.

(Continued)

Table 6-15. (*Continued*)

Test method	Purpose of test	Principle of operation	Advantages	Limitations
Rotational–lateral stiffness method	Test the rotational–lateral stiffness of beam-to-panel assemblies.	The free flange of the beam in the beam-to-panel assembly is loaded parallel to the beam, and the deflections and rotations are recorded.	The test can be used to establish limits of displacements for avoiding joint failures.	The test specimens must be identical to correlate the results.
Structural performance of assemblies and systems	Test configurations of structural elements, assemblies, connections, or details when a safe load-carrying capacity cannot be determined in accordance with the AISI and ASTM specifications.	The assemblies must support, without excessive deflection or failure, the appropriate loads from AISI F1(b).	Tests may reduce the amount of repair costs if the results satisfy the required capacities.	The tests must be performed with the same conditions as those in the building.

(*Continued*)

Table 6-15. (*Continued*)

Test method	Purpose of test	Principle of operation	Advantages	Limitations
Stub-column test method	The test determines the effective cross-sectional area of cold-formed steel columns.	The column is loaded until failure. The failure load and type are recorded. In addition, deflections can be recorded for the incremented loadings.	The test is used to determine the effective area and the ultimate strength of a column. Moreover, the test is used as a means to observe local buckling deformations when the appearance of a column under stress must be observed.	If local buckling is involved, an engineer should be present to observe the failure locations and modes.
Test method for composite slabs	The test results shall be used to determine the strength and behavior of composite steel deck slabs.	The test specimen is loaded until failure. The load, failure type, and deflections are recorded.	The test results on mode failure can be obtained. The failure should be flexure-bond and not shear-bond.	The test specimen must be prepared to simulate, as closely as possible, the field conditions.

Table 6-16. Evaluation of Properties of Structural Steel.

See Section 6.2.13.2 Tables	Table 6-12	Table 6-12	Table 6-12	Table 6-12	Tables 6-12 and 6-13	Table 6-12	Table 6-12	Table 6-12
Evaluation procedure / Chemical and physical properties	Bend tests	Chemical analysis	Fatigue tests	Fracture tests	Hardness tests	Impact test	Modulus test	Tension test
Carbon content		•						
Chemical composition		•						
Ductility	•							
Elongation								•
Fatigue properties								
Cycle counting			•					
Endurance limit			•					
e–N curve			•					
Hardness					•			
S–N curve			•					

(*Continued*)

Table 6-16. (*Continued*)

See Section 6.2.13.2 Tables	Table 6-12	Table 6-12	Table 6-12	Table 6-12	Tables 6-12 and 6-13	Table 6-12	Table 6-12	Table 6-12
Evaluation procedure / Chemical and physical properties	Bend tests	Chemical analysis	Fatigue tests	Fracture tests	Hardness tests	Impact test	Modulus test	Tension test
Fracture properties								
Charpy number						•		
Fracture toughness				•				
Nil ductility trans. Temp.				•				
Modulus of elasticity							•	•
Reduction in area								•
Tensile strength								•
Yield strength								•

Table 6-17. Evaluation of Physical Conditions of Structural Steel

See Section 6.2 and Table 6-14	Table 6-14	Table 6-14	Table 6-14	Table 6-14	Table 6-14	Table 6-14	Table 6-14	Table 6-14	Table 6-14	Table 6-14
Evaluation procedure / Physical condition	Acoustic emission	Eddy current	Hole drilling	Liquid penetration	Magnetic particle	Physical measurement	Radiography	Structural analysis	Ultrasonic	Visual examination
Bracing of compression Elements and members						•				•
Cross-sectional properties						•				
Deformations						•				•
Direct chemical attack						•				
Electrolytic or electrochemical Corrosion						•				•
Fatigue cracking	•	•		•	•	•	•		•	
Fire protection						•				•

(Continued)

Table 6-17. (*Continued*)

See Section 6.2 and Table 6-14	Table 6-14	Table 6-14	Table 6-14	Table 6-14	Table 6-14	Table 6-14	Table 6-14	Table 6-14	Table 6-14	Table 6-14
Evaluation procedure / Physical condition	Acoustic emission	Eddy current	Hole drilling	Liquid penetration	Magnetic particle	Physical measurement	Radiography	Structural analysis	Ultrasonic	Visual examination
Fracture cracking	•	•		•	•	•	•		•	•
Geometry of structural components						•				•
In situ stresses			•							
Laminar tearing	•	•		•	•	•	•		•	•
Overall or local buckling						•		•		•
Overstressing								•		

Table 6-18. Evaluation of Properties of Connectors of Metals.

See Section 6.2 and Tables	Table 6-12	Table 6-5	Table 6-12	Table 6-12	Table 6-5
Evaluation procedure / Chemical and physical properties	Chemical analysis	Coating test	Fastener mechanical tests	Steel product mechanical tests	Stress extension tests
Bolts and rods (including washers and nuts)					
Chemical composition	•				
Hardness			•		
Proof load			•		
Tensile strength			•		
Rivets					
Chemical composition	•				
Hardness			•		
Studs					
Chemical composition	•				
Elongation			•		
Reduction in area			•		
Tensile strength			•		

(Continued)

Table 6-18. (*Continued*)

See Section 6.2 and Tables	Table 6-12	Table 6-5	Table 6-12	Table 6-12	Table 6-5
Evaluation procedure / Chemical and physical properties	Chemical analysis	Coating test	Fastener mechanical tests	Steel product mechanical tests	Stress extension tests
Welds					
Tensile strength of filler material				•	
Yield strength of filler material				•	
Wire strand and wire rope					
Breaking strength				•	
Coating weight		•			•
Elongation				•	
Stress at 0.7% extension					•
Tensile strength				•	•

Table 6-19. Evaluation of Physical Conditions of Connectors of Metals.

See Section 6.2 and Table 6-14	Table 6-14	Table 6-14	Table 6-14	Table 6-14	Table 6-14	Table 6-14	Table 6-14	Table 6-14	Table 6-14
Evaluation procedure / Physical condition	Acoustic emission	Bolt tightness	Eddy current	Liquid penetration	Magnetic particle	Physical measurements	Radiographic	Ultrasonic	Visual examination
Bolts and rods (including washers and nuts)									
Condition						•			•
Corrosion						•			•
Cross-sectional properties						•			
Deformation						•			•!
Dimensions						•		•	
Nonbinding condition						•			
Notched holes for Movement						•			
Tightness		•							

(Continued)

Table 6-19. (*Continued*)

See Section 6.2 and Table 6-14	Table 6-14	Table 6-14	Table 6-14	Table 6-14	Table 6-14	Table 6-14	Table 6-14	Table 6-14	Table 6-14
Evaluation procedure / Physical condition	Acoustic emission	Bolt tightness	Eddy current	Liquid penetration	Magnetic particle	Physical measurements	Radiographic	Ultrasonic	Visual examination
Rivets									
Condition						•			•
Corrosion						•			•
Cross-sectional properties						•			
Deformation						•			•
Dimensions						•			
Soundness									•
Tightness									•
Studs									
Condition						•			•
Corrosion						•			•
Cross-sectional properties						•			
Deformation						•			•

(*Continued*)

Table 6-19. (*Continued*)

See Section 6.2 and Table 6-14	Table 6-14	Table 6-14	Table 6-14	Table 6-14	Table 6-14	Table 6-14	Table 6-14	Table 6-14	Table 6-14
Evaluation procedure / Physical condition	Acoustic emission	Bolt tightness	Eddy current	Liquid penetration	Magnetic particle	Physical measurements	Radiographic	Ultrasonic	Visual examination
Welds									
Cracks	•		•	•	•		•	•	
Discontinuities	•		•	•	•		•	•	
Length						•			
Location						•			•
Profile						•			•
Porosity	•		•	•	•		•	•	
Size						•			•
Slag deposits	•		•	•	•		•	•	
Smoothness									•
Uniformity									•
Wire strand and wire rope									
Condition						•			•
Corrosion						•			•
Cross-sectional properties						•			
Length						•			
Tightness									•

6.3 CONDITION ASSESSMENT OF MASONRY

The condition of a masonry structure is best determined through an understanding of its material properties and physical conditions by visual examination in conjunction with applicable computational methods and testing. Major material properties and physical conditions are discussed in Section 6.3.4. Test methods that can be used to evaluate these properties and physical conditions are provided in Section 6.3.13. Proper assessment and subsequent evaluation (Section 6.3.1) should be utilized to reveal the masonry's ability to transmit and sustain loads and endure the impacts of the environmental conditions to which it is being or will be subjected.

Reference is made to Sections 6.1 (Concrete) and 6.2 (Metals) for the evaluation of reinforcing steel bars, masonry ties, anchors, and joint reinforcement. For strength of anchors and ties in masonry, see Table 6-21, ASTM E488, and ASTM E754. For the evaluation of concrete, metal, and wood used with masonry, reference is made to Sections 6.1, 6.2, and 6.4, respectively.

An engineering analysis of existing masonry structures requires a knowledge of how new structures are designed and how masonry structures behave when they are stressed or displaced. TMS 402/602, *Building Code Requirements and Specification for Masonry Structures* (TMS 2022), should be referenced for this purpose. Older existing structures may require the following references: earlier editions of TMS 402/602 (1988 to present); *Brick and Tile Engineering: Handbook of Design* by the Structural Clay Products Institute (Plumber 1950): *Principles of Tile Engineering and Principles of Brick Engineering* and *Recommended Practice for Engineered Brick Masonry* by the Structural Clay Products Institute (1969). Additional information regarding an assessment of masonry structures can be found in TMS 1700 (2017), *Guide for Condition Assessment of Masonry Structures*, ASTM C1496, *Standard Guide for Assessment and Maintenance of Exterior Dimension Stone Masonry Walls and Façades* (2018), and ICRI 410.1, *Guide for the Evaluation of Masonry Facade Structures* (2008).

6.3.1 Reasons for Conducting a Condition Assessment of Masonry

A structural condition assessment may be conducted on a masonry structure for a number of reasons; the most common *technical* motivations for an assessment often relate to one or more of the following categories:

Material Properties' Questions: Knowing the masonry material properties is necessary for a better understanding of its durability, density, and behavior for preventing deterioration; for an analysis of new or modified loads; and to analyze modifications as needed to best extend service life.

Moisture Issues: Determining moisture levels is necessary to assess the impact of moisture content and to determine the sources of the moisture to better understand the masonry's durability, reasons for deterioration, and gauge future performance.

Observed Deterioration: Comprehensive survey of observable deterioration is needed to determine the cause, locate, and quantify the extent, and assess its impact on material properties. Moisture is a major cause of deterioration in masonry structures.

Strength or Stiffness Questions: There is a need to determine the strength and stiffness of individual members or components to understand their capacity and mechanical properties and evaluate their adequacy for existing or proposed loads.

Structural Capacity and Performance: There is a need to verify by calculation or other means the ability of the masonry unit or assembly to carry load and perform as intended and required.

Environmental Damage: There is a need to understand the effects of wind, earthquakes, fire, and so on, on masonry units and assemblies.

Note that over time, there has been a greater emphasis on energy efficiency of buildings. For existing buildings, addressing energy efficiency is beyond the scope of a typical masonry condition assessment; however, modifications to existing buildings to make them more energy-efficient, such as adding insulation in exterior walls or at roofs, can alter the moisture and temperature environment. Observation of modifications and/or alterations to the original building may warrant additional consideration when conducting a masonry condition assessment.

6.3.2 Masonry as a Material

In its simplest forms, a masonry assemblage can consist of dry masonry units or masonry units and mortar. In complex assemblages, different types of masonry units are used as well as different types of mortar and can also include steel-reinforcing bars, grout, tie rods, embedded metals and wood, metal ties, anchors, and joint reinforcement. In some cases, the masonry assemblage includes an air space, as in a cavity wall. The masonry units may be manufactured products, or they may be dimension stones.

Masonry assemblages include all non-load-bearing walls, load-bearing walls, beams, columns, piers, panels, curtain walls, veneers, partitions, floor arches, arches, and shells made from both masonry and dimension stone masonry units. These units may be bedded or bonded together by mortar and/or grout and may use masonry ties, anchors, tie rods, and joint reinforcement as connectors. Masonry can also be used compositely with concrete and metals and as veneer with concrete, wood, and steel structure backup.

Masonry assemblages commonly consist of two or more components (Section 6.3.4.2) that may act compositely or may be reinforced and, in general, provide multiple functions. Because of their space enclosing nature, they not only serve as structural elements to carry building loads but also provide architectural features and protection from the environment. Accordingly, in addition to evaluation of the properties of masonry, evaluation of the physical conditions can be approximately divided into those conditions that are influenced by the environment and those conditions influenced by the structure. In addition to environmental and structural influences on the physical conditions of masonry, related architectural features of the construction that are not part of the masonry assemblage may also directly influence the physical conditions. Investigation must explore how the total assemblage fits together and how the various components are tied or held together.

For the condition assessment of other materials, reference is made to Section 6.1 for Concrete, Section 6.2 for Metals, and Section 6.4 for Wood. This section addresses the masonry units, mortar, and grout of the masonry assemblage.

6.3.3 Terminology Related to Masonry Characteristics and Properties

Regardless of whether it is an individual masonry unit or part of a masonry assemblage, the basic terminology that is relevant to a masonry condition assessment is the same. Rather than reproduce in their entirety in this book, TMS 402/ACI 530/ASCE 5 2013, TMS 602/ACI 530.1/ASCE 6, TMS 1700, *Guide for Condition Assessment of Masonry Structures*; ASTM C1496, *Standard Guide for Assessment and Maintenance of Exterior Dimension Stone Masonry Walls and Façades*; and ICRI 410.1, *Guide for the Evaluation of Masonry Façade Structures*, a few definitions that can be of assistance in the assessment of masonry have been included here.

Absorption: Process by which liquid is drawn into and tends to fill permeable pores in a porous solid form. The amount of water that a masonry unit (solid or hollow clay or concrete; or dimension stone) absorbs when immersed in either cold or boiling water for a stated length of time. It may be expressed as percentage of mass (clay masonry) or mass per unit volume (concrete masonry).

Adhesion: Property denoting the ability of a material to resist delamination or separation into two or more layers. A property of prefaced masonry units. When viewed without magnification, no visible failure of the adhesion of the facing material to the masonry unit shall occur after the unit has been subjected to a standard compressive strength test.

Autoclave Crazing: Autoclave crazing is failure of the ceramic glazing on masonry units (structural clay facing tile, facing brick, and solid masonry units) when subjected to a standard steam pressure in an autoclave for a specified period.

Bulk-Specific Gravity: Ratio of the density of a material to the density of some standard material, such as water, at a specific temperature. A property of dimension stone equal to the mass of an oven-dried specimen divided by the difference between the mass of the soaked and surface-dried specimen in air and the mass of the soaked specimen in water.

Chemical Resistance: Ability to resist chemical attack. A property of the ceramic glazing on masonry units (structural clay facing tile, facing brick, solid masonry unit, and chemical resistant units) indicating its resistance to changes in color and texture when subjected to immersion in a specified acid solution for a specified length of time and temperature.

Compressive Strength: Maximum compressive stress a material can withstand without failure. For a masonry unit, the maximum load divided by the gross area for brick units and the gross or net area for concrete masonry units as defined by the referenced standards when subjected to a standard strength test.

Dimensions: Nominal measurements of width, height, length, face shell thickness, and web thickness of masonry units.

Distortion: Change in the shape of a solid body similar to warpage. Specified tolerances are required for hollow brick, facing brick, and ceramic glazed masonry units (see **Warpage**).

Drying Shrinkage: Change in the linear dimension of a test specimen caused by drying from a saturated condition to an equilibrium weight and length under specified accelerated drying conditions. Applicable to concrete masonry units.

Efflorescence: Deposit of soluble salts, in general, white and most commonly consisting of sulfate and chloride, that may form on the surface of a masonry unit when moisture moves through the unit followed by precipitation or crystallization at the surface. Masonry units are rated as "not effloresced" and "effloresced."

Freeze–Thaw Resistance: Resistance of masonry units to the cyclical action of temperatures dropping below or rising above the freezing point of water.

Imperviousness: Capability of masonry units to resist being permeated by gases and liquids. The resistance to staining of ceramic glazed masonry units when subjected to a standard test.

Initial Rate of Absorption (IRA): Measure of the suction of water upward into a dry brick or structural clay unit from a bed face partially immersed in water for one minute, expressed in kg/m^2/min (also sometimes called *suction*). Flashed brick may exhibit significantly different IRA values from one bed surface to the other if one surface is flashed and the other is not. Molded bricks definitely have different IRA characteristics among bed surfaces (Bailey et al. 1990).

Mass Determination: Mass per unit area of a specimen is calculated by dividing the total mass, determined after a specified procedure of oven-drying and cooling of a unit, by the average projected area of the two faces of the unit as normally laid in a wall.

Modulus of Elasticity: Ratio, within the elastic limit of a material, of stress to corresponding strain under given loading conditions.

Modulus of Rupture: Value of the maximum tensile or compressive stress (whichever causes failure) in extreme fiber of a unit loaded to failure in bending computed from the flexure formula.

Moisture Content: Amount of moisture in a material determined under prescribed conditions and expressed as a percentage of the mass of the moist specimen; that is, the original mass comprising the dry substance plus any moisture present. For concrete masonry units, it is the percentage of total absorption.

Moisture Expansion: Increase in the dimension or bulk volume of some masonry units caused by reaction with water or water vapor. This reaction may occur in time at atmospheric temperature and pressure but is expedited by exposure to water or water vapor at elevated temperatures and pressures.

Net Area: Net area of concrete masonry units is defined as the ratio of the net volume to the height of the unit. Net area calculations are based on values obtained in absorption and unit mass tests.

Opacity: Degree of obstruction to the transmission of visible light. A requirement for certain types of ceramic glazed masonry and glass block units.

Sampling: Method of selecting the amount of masonry units that are representative of the whole for testing.

Saturation Coefficient: For brick masonry and structural clay tiles, the ratio of the saturated mass of a specimen (after 24 h submersion in cold water) minus the dry mass of the specimen to the saturated mass of a specimen (after 5 h′ submersion in boiling water) minus the dry mass of the specimen. An indication of the probable resistance of brick to freezing and thawing.

Secant Modulus of Elasticity: Slope of the line connecting the origin and a given point on the stress–strain curve.

Size: Measurement of width, depth, and length for masonry units (refer to **Dimensions**).

Solubility: Extent that one material will dissolve in another typically expressed as mass or volume percent at a certain temperature. The loss in mass as a percentage of original mass as determined by a test for solubility in sulfuric acid. A requirement for chemical-resistant masonry units.

Splitting Tensile Strength: Failure load of a test specimen of a masonry unit when subjected to a diametric compressive load along its length until failure. Used as a measure of diagonal tension capacity and of tensile strength. The tensile strength of masonry units is an important characteristic affecting the capacity of concrete block masonry under various loading conditions. The tensile strength of the block is highly sensitive to the test technique. The strain gradient has a significant effect on block tensile strength (Hamid and Drysdale 1980).

Surface Texture: Geometric irregularities in the surface of a masonry unit. Measurement of surface texture shall not include inherent structural irregularities unless these are the characteristics being measured.

Thermal Expansion: Increase in the dimensions or volume of a body caused by a change in temperature. The coefficient of thermal expansion is the change in unit length (or volume) accompanying a unit change in temperature at a specified temperature.

Void Area Ratio in Cored Units: Masonry units' ratio of void area to the solid area.

Warping: Curvature of a flat surface measured as the deviation of the surface from a true plane along edges or diagonals. The maximum out-of-plane convexity or concavity shall be measured to the nearest 1/32 in. (1 mm).

Water Absorption: Increase in weight of a test specimen after immersion in water under specified conditions of time and temperature, expressed as a percentage of its dry weight.

Young's Modulus: Ratio of normal stress to corresponding strain for tensile or compressive stresses less than the proportional limit of the material.

6.3.4 Physical and Mechanical Properties of Masonry

Properties of masonry have been separated into Properties of Masonry Units and Properties of Masonry Assemblages. However, physical conditions have not been separated because the physical conditions apply to both. Section 6.3 follows these approaches. Tables 6-20 and 6-21 combine tests on Masonry Units and Masonry Assemblages. However, the application of each test, either to units or assemblages, is indicated in the application column where appropriate.

6.3.4.1 Properties of Mortar and Grout. Mortar is a mixture of cementitious materials, aggregates, possible admixtures, and water which, in its plastic state, is suitable for laying and bonding masonry units. Grout is a generic term reflecting a family of mortar and concrete mixtures usually used for filling voids or cavities in masonry, which may contain steel reinforcing (Section 6.1). Grout usually increases the load resistance of masonry or masonry assemblages.

Samples of mortar and grout taken from structures are usually insufficient in size for compressive strength and splitting tensile strength testing. All physical properties of mortar and grout must be obtained through a petrographic analysis. The properties of mortar and grout include the following:

Aggregate Characteristics: Visual examination can determine the composition, fabric, and condition of the aggregate. Microscopic examination can determine the size, shape, grading, distribution, texture, composition, rock types, alteration (degree and products), coating, rims, internal cracking, and contamination.

Constituent Portions: Proportions of cementitious materials/lime/aggregate or other constituents in the mortar or grout. Proportions can be determined with chemical and other analysis techniques.

Mortar and Grout Characteristics: Visual examination can determine cracking, color change, embedded items, and condition. Microscopic examination can determine the existence of air entrainment, air voids (size, shape, and distribution), bleeding, segregation, aggregate-paste bond, fractures, alteration (degree, type, and reaction products), and nature and condition of surface treatments.

Paste Characteristics: Microscopic examination can determine color, hardness, porosity, carbonation, residual cement, mineral admixtures, compounds in hydrated cement, and contamination. This examination provides information regarding distribution, particle size, abundance, and mineral admixture identification.

- Primary functions of masonry mortar are to provide for even bearing, bond the masonry units together, and provide resistance to water penetration (Hatzinikolas et al. 1980).
- Grouted masonry testing showing grout compressive strength, per se, is a poor indicator of the interaction of grout in masonry (Isberner 1980).

6.3.4.2 Properties of Masonry Assemblages. Masonry assemblages may be cut from a structure and tested as prisms or panels. Strength and stress–strain characteristics of masonry prisms made with similar units differ markedly depending on the direction of compression and whether the prisms are hollow, solid, or grouted (Wong and Drysdale 1983). The properties of masonry assemblages include the following:

Compressive Strength: Compressive force imposed on a masonry prism at any given moment divided by the masonry prism area.

Concentrated Load Capacity: Ability to support superimposed point loads.

Deformability: Measurement of masonry assemblage deformability properties by use of testing equipment such as thin, bladder-like flat jack devices installed in saw-cut mortar joints in the masonry and measuring load–deformation (stress–strain) properties.

Flexural Tensile Strength, also known as Flexural Bond Strength: Determination of the strength of unreinforced masonry assemblages based on tests of simply supported beams (bond wrench test).

Modulus of Rupture: Value of the maximum tensile stress in the extreme fiber of a specimen loaded to failure in bending computed from the flexure formula.

Secant Modulus of Elasticity: Slope of the line connecting the origin and a given point on the stress–strain curve. A property obtained in a strength test on a masonry prism.

Shear Strength: Maximum shear or diagonal tensile stress of masonry assemblages loaded in compression along one diagonal, thus causing a diagonal tension failure with the specimen splitting apart parallel to the direction of load.

Tensile Load Capacity: Maximum load that a masonry panel can sustain when subjected to a standard tensile test.

Transverse Load Capacity: Maximum load that a masonry panel can sustain when subjected to a standard transverse load test. The load is applied as two-line loads located at the quarter points of the span being loaded.

6.3.4.3 Physical Conditions of Masonry Units and Masonry Assemblages (Environmental). The physical conditions of masonry that are influenced by environmental conditions (deicing salts, freezing and thawing, moisture, thermal changes, and weathering) include the following:

Air Leakage: Volume of air per minute at a measured temperature and pressure that flows perpendicularly through a masonry assemblage.

Blistering: Swelling accompanied by rupturing of a thin uniform skin both across and parallel to the bedding plane, usually a condition found on sandstone but also on granite.

Calcite Streak: Former fracture or parting (in limestone) that has been re-cemented and annealed by the deposition of white or light-colored calcite.

Coving: Hollowing out of a wall just above grade level, usually caused by standing rainwater or rainwater splash off the ground.

Cracking: Fissures in masonry 0.004 in. (0.1 mm) or larger in width (Figure 6-24).

Crazing: Formation of a pattern of tiny cracks in the glaze of glazed masonry units (structural clay facing tile, facing brick, and solid masonry units).

Crumbling: Condition indicative of a certain brittleness or tendency of the masonry to break up or dissolve. *See* **Friability**.

CryptoFlorescence: Potentially harmful accumulation of soluble salts crystalized under or just beneath the masonry surface as moisture in the masonry evaporates.

Delamination: Condition of masonry in which the outer surface splits apart into laminae or thin layers but remains in place (Figure 6-25).

Deterioration: Degradation of masonry resulting in a permanent impairment of the physical properties (Figure 6-26).

Dew Point: Temperature at which air becomes saturated with water vapor at the existing pressure and below which moisture is likely to condense; varies with the amount of moisture in the air.

Figure 6-24. Step cracking within a brick masonry veneer.
Source: Courtesy of Raths, Raths & Johnson, Inc.

Figure 6-25. Face brick delamination.
Source: Courtesy of Raths, Raths & Johnson, Inc.

Driving Rain Index: Measure of the likelihood of rain penetration determined by combining the annual precipitation with the average annual wind velocity.

Efflorescence: Deposit or encrustation of soluble salts, typically white and most commonly consisting of sulfates and chloride, that may form on the surface of masonry when moisture moves through the assemblage followed by evaporation and crystallization (Figure 6-27).

Erosion: Wearing away of the surface, edges, corners of carved details of masonry slowly and usually by external actions of chemicals or the mutual action of windblown particles and water.

Exfoliation: Natural stone deterioration such as peeling, scaling, or flaking of the surface of stone in thin layers.

Friability: Inherent characteristic of some types of stone, particularly sandstone or limestone, which have a tendency to break up, crumble, or powder easily.

Interstitial Condensation: Accumulation of moisture that occurs within a masonry assemblage when moist air reaches its dew point.

Peeling: Debonding of the glaze surface or coating of masonry caused by a lack of adhesion or improper application of a surface treatment.

Pitting: Small cavities in a masonry surface as a result of natural weathering or harsh or abrasive cleaning.

Rising Damp: Suction of groundwater into the base of masonry walls through capillary action (Figure 6-28).

Figure 6-26. Face brick freeze–thaw deterioration.
Source: Courtesy of Raths, Raths & Johnson, Inc.

Figure 6-27. Mass wall masonry interior face efflorescence (white deposits).
Source: Courtesy of Raths, Raths & Johnson, Inc.

Salt Fretting: Condition of masonry resulting in an obvious pattern of erosion caused by salts.

Spalling: Condition of masonry in which the outer layer or layers of masonry units break off (unevenly) or peel away in parallel layers from the larger block of masonry (Figure 6-29).

Figure 6-28. Exterior limestone and a brick masonry rising damp.
Source: Courtesy of Raths, Raths & Johnson, Inc.

Staining: Discoloration of masonry arising from foreign materials.

Sugaring: Characteristic of some masonry indicative of gradual surface disintegration (i.e., a granular or powdery condition).

Surface Condensation: Accumulation of moisture that occurs on the surface of masonry when moist air reaches its dew point.

Surface Deposits/Surface Induration: Accumulation of products on the masonry surface to a measurable thickness caused by this movement or moisture.

Surface Temperature: Physical temperature of the masonry at its surface as distinguished from air temperature.

Tidemark: Stain on a masonry wall that occurs where equilibrium has been reached between capillarity and evaporation.

Transition Area: Area above the tidemark on a masonry wall that reflects a changing tidemark as equilibrium conditions vary with the weather.

Warping: Curvature of a flat surface measured as a deviation of the surface from a true plane along the edges or the diagonals of the unit. The maximum out-of-plane convexity or concavity shall be measured to the nearest 1/32 in. (1 mm).

Water Penetration/Permeance: Physical passage of water through a masonry assemblage by means of interconnected pores (measured by the permeability of the materials) or through cracks and other physical openings.

Weathering: Natural disintegration and erosion of masonry caused by environmental conditions, resulting in granular and rounded surfaces.

6.3.4.4 Physical Conditions of Masonry (Structural). The physical conditions of masonry that are influenced by structural conditions (movement of support, insert slippage, settlement, overloads, volume changes in the structure owing to elastic shortening, creep, thermal changes, etc.) include the following:

Bowing: To displace a portion of masonry assemblage into a curve.
Bulging: To displace a portion of a masonry assemblage outward.
Chipping: Cutting or breaking of small pieces from masonry units.
Cracks or Cracking: To break, split, or snap apart in masonry resulting in narrow fissures of 0.004 in. (0.01 mm) or larger in width, which occur because of a variety of causes (Figure 6-30).
Deflection: Vertical or horizontal movement of masonry assemblages and structural members or elements.
Detach or Detachment: Separation and a partial or complete break of masonry units or assemblages.
Displacement: Permanent vertical or horizontal movement of all or a portion of masonry assemblages (Figure 6-31).
Distortion: Permanent change in the shape of masonry units or assemblages, including bowing, bulging, sagging, and warping.
Plumbness: Vertical out-of-plane deviation of all or a part of masonry assemblages.
Sagging: Vertical deflection of structural members in a masonry assemblage.
Settlement: Vertical movement of foundations typically resulting in damage to a masonry assemblage.

Figure 6-29. Face brick spalls.
Source: Courtesy of Raths, Raths & Johnson, Inc.

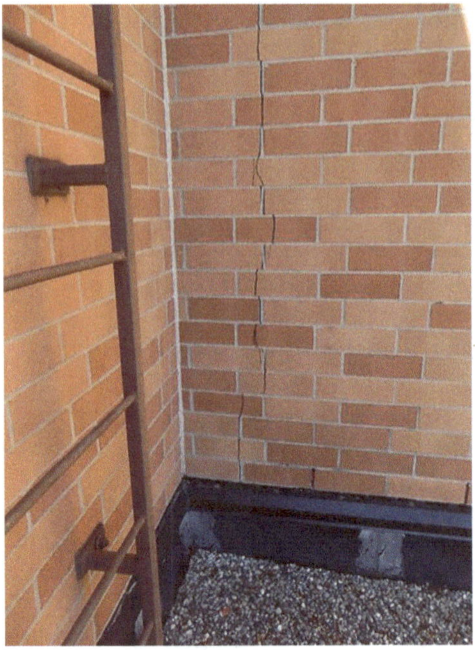

Figure 6-30. Vertical cracking of masonry adjacent to an interior corner.
Source: Courtesy of Raths, Raths & Johnson, Inc.

Shrinkage: Reduction in the dimensions of masonry units or assemblages caused by a decrease in moisture content.

Spalling: Condition in which the outer surface or outer layer of masonry units break off or peel off, usually in parallel layers.

Swelling: Increase in the dimensions of masonry units or assemblages caused by an increase in moisture content. Clay bricks absorb water or moisture and expand with time.

Warping: Permanent out-of-plane distortion of portions of masonry assemblages.

6.3.4.5 Physical Conditions of Masonry (Architectural). The physical conditions of associated architectural features and details that have a significant influence on the physical conditions of masonry include the following:

Condition of Flashing: Functional state of the integrity of the thin impervious material placed at interruptions in the masonry to either prevent water infiltration or control after infiltration and redirect it to the exterior.

Condition of Roofing: Functional state of the integrity of the material or assembly that serves as the roof covering to waterproof the top of the building.

Condition of Control Joints: Joints between adjacent parts of masonry that accommodate contraction of the adjacent masonry, thereby avoiding the development of stresses high enough to cause cracking in the masonry (Figure 6-32).

Condition of a Damp Course: Impervious horizontal layer of material in a masonry assemblage used to prevent the capillary flow of moisture, usually from the ground, but also used to prevent seepage from above.

Condition of Dampproofing: Treatment to the surface of a masonry assemblage used to resist the passage of water in the absence of hydrostatic pressure.

Figure 6-31. Horizontal displacement of masonry at an exterior corner; movement up to 1.75 in.
Source: Courtesy of Raths, Raths & Johnson, Inc.

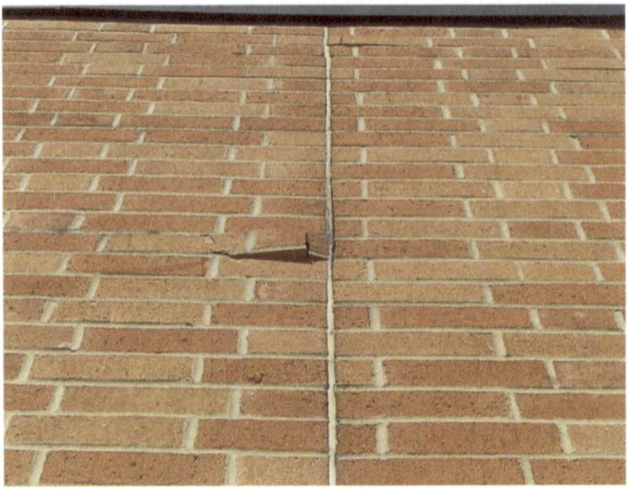

Figure 6-32. Face brick crack/spall at a poorly installed masonry control joint. Source: Courtesy of Raths, Raths & Johnson, Inc.

Condition of Expansion Joints: Joints between adjacent parts of masonry that accommodate movements caused by changes in temperature, moisture expansion, elastic deformation, settlement, and creep (BIA Technical Note 18a).

Location: Position of a portion of the masonry assemblage in relation to the entire building structure. The location may have a significant influence on physical conditions.

6.3.4.6 Physical Conditions of Mortar and Grout. The physical conditions of mortar and grout include the following:

Air Content: Volume of air voids in the mortar or grout, usually expressed as a percentage of the total volume.

Alignment: Lines that establish the position of construction of a masonry assemblage.

Bed Joint Levelness: Levelness or deviation from a horizontal plane of the layer of mortar on which masonry units are set.

Bed Width Variation: Variation in thickness of the layer of mortar on which masonry units are set.

Condition: A general overall assessment of the mortar or grout and its ability to function as intended.

Cracking: Fissures 0.004 in. (0.1 mm) or larger in width in, or adjacent to, the mortar or grout.

Exposure of Embedded Metal Components: Visible exposure of masonry ties, anchors, and reinforcement provides an opportunity for rapid corrosion of the component.

Friability: Mortar or grout that is easily crumbled or pulverized and reduced to powder.

Hardness: Term relating to the capacity of mortar or grout to withstand permanent deformation.

Pitting: Development of small cavities in the surface of grout and mortar owing to phenomena such as cavitation or local disintegration.

Sandiness: Presence of sand on the exposed surfaces of mortar and grout indicative of cement deficiency or excessive acid washing.

Voids: Air spaces within mortar or grout or between masonry units and the mortar or grout.

6.3.5 Components of Masonry Assemblages

6.3.5.1 Manufactured Masonry Units. Manufactured masonry units include the following (Section 6.3.13) (Table 6-20 for references):

Structural Clay Load-Bearing Wall Tile (ASTM C34);
Concrete Brick (ASTM C55);
Structural Clay Non-Load-Bearing Tile (ASTM C56);
Building Brick (Solid Masonry Units Made from Clay or Shale) (ASTM C62);
Calcium Silicate Face Brick (Sand-Lime Brick) (ASTM C73);
Load-Bearing Concrete Masonry Units (ASTM C90);
Ceramic Glazed Structural Clay Facing Tile, Facing Brick, and Solid Masonry Units. (ASTM C126);
Non-Load Bearing Concrete Masonry Units (ASTM C129);
Structural Clay Facing Tile (ASTM C212);
Facing Brick (Solid Masonry Units Made from Clay or Shale) (ASTM C216);
Chemical-Resistant Masonry Units (ASTM C279);
Hollow Brick (Hollow Masonry Units Made from Clay or Shale) (ASTM C652);
Prefabricated Masonry Panels (ASTM C901);
Glass Block Wall (ASTM C240);
Adhered Brick Masonry (TMS 402); and
Adhered Stone Masonry (TMS 602).

6.3.5.2 Natural Dimension Stones. Natural dimension stone masonry units include the following (Section 6.3.13, Table 6-20 for references):

Marble Dimension Stone (Exterior) (ASTM C503),
Limestone Dimension Stone (ASTM C568),
Granite Dimension Stone (ASTM C615),
Quartz-Based Dimension Stone (ASTM C616), and
Slate Dimension Stone (ASTM C629).

6.3.5.3 Mortar and Grout. Mortar and grout used in masonry include the following: (1) *Mortar for Unit Masonry* (ASTM C270), and (2) *Grout for Masonry* (ASTM C476).

6.3.5.4 Reinforcing Steel Bars. For reinforcing steel bars used in masonry, see Section 6.1.

6.3.5.5 Masonry Ties, Anchors, and Joint Reinforcement. Masonry ties, anchors, and joint reinforcement are fabricated from metals, and reference is made to Section 6.2.

6.3.5.6 Connections. Typically, mortar is used to bed or bond masonry units on or to each other. When mechanical connections are used, they are typically made of metals, and reference is made to Sections 6.2 and 6.3.8 for their condition assessment.

6.3.5.7 Other Materials. Masonry assemblages can include concrete, metal, and wood elements. Reference is made to Sections 6.1, 6.2, and 6.4, respectively, for these materials.

6.3.6 Mechanisms of Masonry Deterioration

Masonry includes not only manufactured units such as brick and concrete masonry units but also dimension stone and mortar. The physical conditions that masonry is subjected to during its service life are the primary causes of deterioration. For convenience, these conditions have been grouped into environmental attack, structural deficiencies or incompatibilities, inadequate architectural features, and mortar and grout deterioration (Grimm 1985, Raths 1985). For causes of deterioration of reinforcing steel in reinforced masonry, reference is made to Sections 6.1 and 6.2.

6.3.6.1 Environmental Attack. Environmental attack that causes deterioration of masonry includes chemical attack, freezing and thawing, moisture, and weathering. Some of these conditions occur together. Driving rain can not only introduce salts into the masonry, which results in salt fretting of stone (a pattern of erosion caused by salts, usually from the salting of sidewalks and pavements), but also contribute to water penetration and erosion of the surface, edges, and corners of stone masonry. Moisture or water penetration into the interconnected pores (measured by the permeability of the masonry) and through cracks and other physical openings can result in efflorescence, cryptoflorescence, surface deposits, and interstitial condensation. Moisture and chemical penetration into masonry can lead to disintegrating cycles of freezing and thawing and corrosion of masonry ties, anchors, joint reinforcement, and reinforcing steel. Freezing and thawing, weathering, and corrosion of embedded metal result in cracking, delamination, erosion, exfoliation, and friability of natural stones, pitting of the masonry surface, and spalling.

6.3.6.2 Structural Incompatibility. Deterioration of masonry can be caused by structural incompatibility, including movement of supports, insert slippage, settlement of foundations, overloads, and volume changes in the structure caused by elastic shortening, creep, or thermal changes. Masonry is often used as a finish material in conjunction with structural reinforced concrete or structural steel. The finish masonry is dependent on its attachment and support from the structure. These supports can move (vertical, horizontal, or rotational movement of supporting shelf angles), and inserts supporting shelf angles can slip. The supporting structure can settle (perhaps on account of overloads) and undergo volume changes. These movements can result in bowing and bulging, chipping, cracking, detachment, displacement, distortion, sagging, spalling, and warping of the masonry.

6.3.6.3 Inadequate Architectural Features. Deterioration of masonry can result from inadequate architectural features. These include the condition of the flashing (loss of integrity permitting entrance of moisture); inadequate vertical joints to control and accommodate contraction and expansion, settlement, and temperature and moisture volume changes; missing damp proof courses used to prevent capillary flow of moisture; clogged weep holes; and lack of horizontal expansion joints where deflection of the supporting members would allow the mass of masonry above a support to be transmitted to the masonry below the support.

6.3.6.4 Mortar and Grout Deterioration. Masonry units may be joined together by mortar and/or grout. Poorly constructed mortar joints can result in misalignment of the masonry units and variation in bed thickness and levelness. Poorly mixed mortar and/or grout can result in cracking of the mortar and grout permitting access for moisture. Improperly constructed masonry joints and improper grouting can also permit exposure of masonry ties, anchors, joint reinforcement, and reinforcing steel to deterioration.

6.3.7 Factors That Influence Masonry Behavior

Deterioration: Freeze–thaw and chemical attack are common causes of masonry deterioration, whether it is mortar or masonry. Microcracking within the masonry or mortar reduces its compressive and tensile strengths and results in a weaker masonry assemblage. Masonry and mortars with high porosity and low strength are more susceptible to freeze–thaw deterioration.

Temperature: High temperatures will affect the strength of the masonry assemblage by reducing the compressive strength when heated above 600 °C (1,112 °F) and also the modulus of elasticity when heated above 200 °C (392 °F).

Moisture: Moisture within the masonry assemblage reduces the compressive strength of the masonry and mortar and also the shear and bond strength of the masonry assemblage.

6.3.8 Connections

The masonry connection condition assessment includes the determination of whether the anchorages of the masonry to the structural support system have been compromised. The initial assessment of the masonry and the anchors or ties is typically visual and performed by someone who is versed in masonry connections. Because the connections are typically metal, corrosion of the anchors, rods, or ties is the typical connection distress.

Common connections include continuous wire ties between backup and masonry veneer, sheet metal anchors between backup and masonry veneer, individual anchors (wire ties) between the backup and the masonry veneer, and tie rods within arched and flat masonry floors. The backup is typically masonry (CMU or brick), cold-formed metal framing, steel framing, concrete framing, or wood stud framing. Design requirements for the masonry anchors and ties are provided within the TMS 402, *Building Code Requirements for Masonry Structures* (TMS 2016) and ASTM C1242, *Standard Guide for Selection, Design and Installation of Dimension Stone Attachment Systems* (ATSM 2022). To fully assess connections, the condition of the masonry, backup, connection, and the configuration and geometry of the connections are required.

6.3.9 Terminology Related to Connector Characteristics and Properties

Connectors for masonry typically include ladder- and truss-type joint reinforcement, sheet metal anchors, wire tie anchors, spiral anchors, single-tail and split-tail anchors, and tie rods. The physical condition of the connectors and the surrounding masonry or stone influences the capacity of the connections. The following terms are relevant to connectors when conducting a structural condition survey.

Condition of the Connector: The physical state as determined by observation and with the use of simple physical techniques such as cleaning and scraping; these can be used to establish the existence of many of the physical conditions included herein.

Corrosion: Oxidation of metal owing to chemical reaction between the metal, moisture, and oxygen in the environment; both the masonry and the metal anchor within the vicinity of the anchor can be degraded by corrosion.

Cross-Sectional Properties: Actual dimensions and other geometric properties of connectors.

Deformation: Elongation, bending, or twisting of anchors, ties, and other connectors of the elastic limit.

Geometry: Actual spacing of connectors within the field of the floor or wall as well as the configuration of the anchor within the masonry and backup material. Some important information includes edge distance from the exterior face, embedment within the backup, and slope within the connector.

6.3.10 Physical and Mechanical Properties of Connectors

Connectors are typically metal (ferrous, galvanized, or stainless). The following is a list of connector properties:

Chemical Composition: Indicators of the properties of ties, anchors, and fasteners can be used to establish corrosion resistance and mechanical properties.

Elongation: Requirement for certain types of connectors is the increase in gauge length in a tension test specimen measured prior to fracture as a percentage of the original length. It is a measure of ductility.

Hardness: For threaded fasteners, sufficient hardness ensures that the surface of the fasteners has the capacity to transmit load by bearing; it is a specified requirement for nuts and washers and sometimes for bolts and rods.

Lateral Resistance: Resistance to lateral movement on connectors that are embedded in masonry.

Proof Load: Specified load without a measured permanent set; applicable to bolts and rods.

Tensile Strength: Stress calculated from the maximum load sustained by a metal connector during a tension test divided by the original cross-sectional area of the specimen.

Yield Strength: Stress at which a significant increase in strain occurs without a corresponding increase in stress.

6.3.11 Mechanisms of Connector Deterioration

Connectors are typically metal and are susceptible to attack by moisture, chemicals, mechanical damage, and fire.

Moisture: Primary mechanism for deterioration of metal connectors is moisture, which can lead to corrosion. Metal connectors can lose their cross section, reducing their capacity. Moisture can affect the properties of masonry, reducing the allowable stresses within the masonry and resulting in a reduction in the connection capacity.

Chemical: Metal connectors can be attacked by some airborne chemicals or chemical additives within the masonry or mortar, resulting in the corrosion of the connector and reduction of the connector capacity.

Mechanical Damage: Mechanical damage for masonry anchors is remote, but if an incident did occur, a reduction of the load-carrying capacity would result. The damage may result from cutting of the masonry, impact to the masonry, or other scenarios.

Fire: Fire can reduce the material properties of the masonry and connectors themselves.

6.3.12 Considerations with Masonry in Historic Construction

Masonry has been used as building material for thousands of years. Masonry building construction has changed over the centuries, but the material is basically manufactured in the same manner.

Historic Structures: Prior to conducting an assessment and any repairs, the historic status of an existing structure should be determined, and the impacts of the repairs on the status of the structure should be evaluated. Regardless of any formally recognized historic status, many structures are recognized as places of regional, cultural, and personal significance. As such, every effort should be made to adhere to the Secretary of the Interior's Standards for the Treatment of Historic Properties, which would require the historic character of a property be retained and preserved and prohibit the replacement of intact or repairable historic materials. Local ordinances and historical societies should also be consulted.

Historic Materials: Many masonry structures contain historic materials such as stone, brick, and terra cotta. A knowledge of the historic construction and materials used will assist the assessment team in evaluating the existing conditions and applying appropriate historic repairs. Most historic structures are constructed with load-bearing masonry walls. These load-bearing walls are predominately brick masonry, but some are stone masonry. Historic brick prior to the mid-1800s was handmade, dried, and fired in kilns on-site. When cities needed fire-resistant materials for structures, remote brick manufacturing at yards became the norm. The compressive strength of the 1800s brick can vary greatly. Load-bearing stone structures were typically religious or governmental buildings because they are considered the most permanent buildings.

6.3.13 Tabulation of Test Methods

6.3.13.1 Test Methods (Preliminary Assessment Chart). The relative value to the assessment professional of visual, nondestructive, and destructive field testing is depicted in Table 6-20 The chart is a quick reference for the applicability of certain test methods to determine masonry and connection conditions and properties. Subsequent tables list the capabilities and limitations of various test methods in greater detail, along with a list of references that have additional information. Condition assessment of structures is dynamic, and the investigator should recognize that some technologies listed may become unavailable as devices are removed from the market or replaced by newer technology.

Visual, nondestructive, and destructive tests are used to establish the properties and physical conditions of both masonry and stone used in structures. Most tests are not suitable for testing masonry in in situ. New test procedures, methodologies, and equipment are developed on an ongoing basis, but most are not commercially available or have gained wide acceptance and should, therefore, be used with a sense of caution. It may be necessary to use a combination of methods for satisfactorily predicting properties of wood. References cited in Section 6.3, unless otherwise noted, are included in Appendixes B1 and B3. Additional references are to be found in Appendixes B2 and B4.

Sample Size: Several ASTM standards exist to guide an investigator in determining sample size if a statistically significant sample is desired. Often a statistically significant sample is not available or economically viable. In this case, it is desired to select a sample without bias. This sample could also be a random sample, but not necessarily, as purely random samples in a failure investigation are often not practical given the constraints of the existing structure. As sample size and distribution are determined, care should be exercised to obtain samples from damaged and undamaged and questionable and unquestionable portions of the structure if applicable.

ASTM E122, *Standard Practice for Calculating Sample Size to Estimate, with a Specified Tolerable Error, the Average for a Characteristic of a Lot or Process* (ASTM 2017), can be followed to determine the number of samples required for a desired level of accuracy. The number of samples depends on

- Error, E, that can be tolerated in the estimate of the measured value;
- Preferred probability, or confidence level, that the measured value is within the limit of error; and
- Estimate of the variation, σ, of the measured value from one unit to another within the population.

Assuming that the measured result is normally distributed and that the population is large, the following equation, based on an assumed 95% confidence interval, can be used to calculate the number of samples, n, for testing.

$$n = \left(\frac{1.96\sigma}{E}\right)^2 \tag{6-3}$$

Table 6-20. Relative Assessment Value for Masonry and Connection Conditions and Properties.

Factor	Water testing	Visual inspection	Remote visual inspection	Metal detector	Infrared thermography	Moisture measurements	Stress waves	Surface hardness	Core drilling	Ground-penetrating radar
Architectural abnormalities		Limited	Limited							
Structural abnormalities		Limited	Limited	Limited						Limited
Water penetration	Yes				Yes					
Measure moisture content					Limited	Yes				
Locate deterioration		Limited	Limited		Limited		Limited	Limited		
Quantify deterioration			Limited				Limited		Limited	
Assess strength							Limited	Limited	Limited	

(Continued)

Table 6-20. (*Continued*)

Factor	Water testing	Visual inspection	Remote visual inspection	Metal detector	Infrared thermography	Moisture measurements	Stress waves	Surface hardness	Core drilling	Ground-penetrating radar
Determine modulus of elasticity							Estimate			
Identify hidden construction defect			Yes		Limited					Limited
Locate and Assess Fasteners		Limited		Limited	Limited					Limited
Mortar and Grout Abnormalities		Limited	Limited							Limited

where

n = Number of samples,
σ = Standard deviation of the measured results, and
E = Allowable error in the measured results.

The factor 1.96, based on a desired 95% confidence level, gives a 0.050 probability that the actual value is not within plus or minus the allowable error, E, of the measured property.

Deterministic sampling describes a situation where a set number of samples are available, or sampling is constrained by any of the factors discussed herein, and only a set number of samples can be taken (access, effects on the element under consideration, budget, etc.). This sampling protocol requires engineering judgment to ensure that a representative estimate of the characteristic under consideration is obtained.

6.3.13.2 Test Methods (Comprehensive Assessment Chart). Testing of specimens from the members of a structure to determine chemical and physical properties can provide useful quantitative data. Although removal of specimens may be destructive, careful selection and handling can ensure that the usefulness of the member is not impaired, and the damaged areas can be satisfactorily repaired. Tests on specimens provide specific, quantitative information without relying on interpretation or analytical results. Consideration should be given to physical testing of specimens over nondestructive testing methods to determine similar information. It is also advantageous to use the results of physical tests in specimens from members as base data to corroborate nondestructive testing results. Most nondestructive methods require highly qualified and skilled personnel, specialized equipment, and considerable analysis and interpretation of data to provide qualitative and quantitative information.

Available test methods are cited in Tables 6-21 through 6-29. The tables are not intended to provide the reader with information and details on how to perform and interpret tests but rather to provide information on test methods that are available, what they can accomplish, advantages and limitations of the tests, and a brief statement about qualifications of test personnel. This information is not exhaustive and is constantly subject to change. Advantages, limitations, and costs can change with time. Accordingly, the references are the most important part of these tables, and care has been taken to provide the most current and in-depth information. Perusal of any specific reference will usually lead to an extensive bibliography on the subject. The supplemental references in Appendixes B2 and B4 contain information that can be pursued for additional knowledge.

Each table in this section is followed by a list of related references. These references are numbered to facilitate referral in the Reference row of the evaluation tables, as well as to the reference column in the test method tables.

Table 6-21. Test Methods for Determining Physical Properties of Masonry Units and Masonry Assemblages on Masonry Specimens.

Property	Purpose of test	Principle of operations	Advantages	Limitations
Absorption	Determination of the percentage of absorption by mass of brick, tile, and concrete masonry units and of natural building stone.	Measuring both dry and saturated mass.	Easy test, low cost.	Specimens must be taken from the structure for testing. Results are affected by mortar clogging pores.
Bulk specific gravity	Determination of the bulk specific gravity of natural building stone.	Weighing a dried specimen, soaked and surface-dried specimen, and soaked specimen.	Easy test, low cost.	Specimens must be taken from the structure for testing.
Compressive strength	Determination of the compressive strength of brick, tile, and concrete masonry units and natural building stone.	Specimens of masonry units or stone are capped and tested in a testing machine using standard procedures.	Easy test, low cost.	Test specimens must be taken from the structure for testing.

(*Continued*)

Table 6-21. (*Continued*)

Property	Purpose of test	Principle of operations	Advantages	Limitations
Compressive strength (in situ)	Compressive strength of masonry, in situ.	In situ tests involve installation of a flat jack device in saw-cut mortar joints.	In situ test possible.	For in situ tests, estimated stress is restricted to the width of the flat jack.
Concentrated load capacity	Determination of the concentrated loads that can be supported by masonry assemblages.	Point load applied to unreinforced solid-unit masonry.	In situ test possible.	Special jigs required. Expensive.
Deformability	Determination of the deformability of masonry assemblages.	In situ test involving installation of a flat jack device in saw-cut mortar joints.	Easy test.	Relatively nondestructive.
Durability (see also freeze–thaw resistance)	Determination of the durability of brick masonry as measured by compressive strength, absorption, and saturation coefficient.	Standard tests for compressive strength, absorption, and saturation coefficient.	Easy tests, low cost.	Specimens must be taken from the structure for testing. Results may be misleading. A freeze–thaw resistance test may provide better results.

(*Continued*)

Table 6-21. (*Continued*)

Property	Purpose of test	Principle of operations	Advantages	Limitations
Efflorescence	Determination of the leaching of soluble salts in brick masonry units.	Specimens are immersed in distilled water and then oven-dried, inspected, and rated.	Easy test, low cost.	False results are possible because units are contaminated in the wall.
Flexure strength	Determination of flexural strength of building stone.	Simple beam testing using quarter-point loading.	Useful because it indicates differences in flexural strength among different building stones.	A fairly long sample (of ten times depth) is required to be taken from the structure.
Flexural tensile strength	Determination of tensile strength that can be supported by masonry assemblages.	Load unreinforced solid-unit masonry in various ways.	In situ test possible.	Special jigs required. Expensive.

(*Continued*)

Table 6-21. (*Continued*)

Property	Purpose of test	Principle of operations	Advantages	Limitations
Freeze–thaw resistance	Determination of the resistance of brick masonry units to rapidly repeated cycles of freezing and thawing.	Test specimens under controlled moisture conditions are subjected to standard cycles of freezing and thawing. Specimens are measured for mass before and after tests and percentage changes are recorded.	Provides useful information.	Specimens must be taken from the structure for testing. Requires 10 weeks to complete the test.
Initial rate of absorption	Determination of the initial rate of absorption of brick masonry units and slate.	Specimens are dried, set in water for 1 minute, and mass-determined. Gain in mass is recorded.	Important information to determine the loss of mixing water in mortar and hence water tightness of joints.	Specimens must be taken from the structure for testing.

(*Continued*)

Table 6-21. (*Continued*)

Property	Purpose of test	Principle of operations	Advantages	Limitations
Modulus of rupture and secant modulus of elasticity	Determination of the modulus of rupture of natural building stone and of brick masonry units and the modulus of rupture and secant modulus of rupture of masonry assemblage.	Specimens are tested as beam with midspan loading and tested to failure. Modulus of rupture calculated from breaking load.	The test indicates differences among various specimens. ASTM E72 and E519 can be adapted to in situ tests.	Specimens must be taken from the structure for testing.
Saturation	Determination of the saturation coefficient of brick masonry units.	Same procedure as for determining absorption except that a saturated mass after 24 h and 5 h boiling is used in the calculation.	Saturation coefficient used as a measure of durability.	Specimens must be taken from the structure for testing. Results may be misleading.

(*Continued*)

Table 6-21. (*Continued*)

Property	Purpose of test	Principle of operations	Advantages	Limitations
Shear strength	Determination of shear strength that can be supported by masonry assemblages.	Load unreinforced solid-unit masonry in various ways.	ASTM E72 and E519 can be adapted to in situ tests. UBC 21-6 is an in situ test.	Special jigs required. Expensive. UBC 21.6 has limited use.
Splitting tensile strength	Determination of the splitting tensile strength of masonry units. Can also be used to detect the presence of defects.	Compressive load is applied to the unit by means of bearing rods, resulting in tensile stress distributed over the height of the unit.	Laboratory method to determine the physical properties of masonry.	The level of precision of the test method is not high. Currently, seldom used to determine physical properties.
Strength of anchors	Determination of the static, tensile, and shear strength of postinstalled and cast-in-place anchorage systems in structural masonry.	Loading plates are located at anchors to be tested, and appropriate testing equipment and testing systems are used to determine the strength of the anchors.	In situ strength values of anchorage determined.	Although specimens are not removed from the structure, tests are destructive.

(*Continued*)

Table 6-21. (*Continued*)

Property	Purpose of test	Principle of operations	Advantages	Limitations
Tensile load	Determination of tensile load that can be supported by masonry assemblages.	Load unreinforced solid unit masonry in various ways.	ASTM E72 can be adapted to the in situ test.	Special jigs required. Expensive.
Transverse load	Determination of transverse load that can be supported by masonry assemblages.	Load unreinforced solid unit masonry in various ways.	ASTM E72 can be adapted to in situ test.	Special jigs required. Expensive.
Water penetration	Determination of resistance to water penetration of masonry assemblages.	Test chambers are placed against specimens with controlled temperature and humidity conditions. Water is applied with increased air pressure. Dampness on the back of the specimen is observed.	Provides useful information on the quality of material coatings, leakage, and workmanship. ASTM E514 intended for laboratory use. ASTM C1601 intended for field use.	Water penetration is significantly affected by air pressure in the test chamber.

Table 6-22. NDT Methods for Determining Properties and Physical Conditions of Masonry Units and Masonry Assemblages.

Property	Purpose of test	Principle of operations	Advantages	Limitations
Acoustic impact (hammer test)	Detect delaminations or disbonds in composite systems; detect voids and cracks in materials, e.g., hammer technique to detect defective masonry units.	Surface of the object is struck with a hammer (usually metallic). The frequency and damping characteristics of the "ringing" can indicate the presence of defects.	Portable; easy to perform the test; electronic measuring equipment not needed for qualitative results.	Geometry and mass of the test object influences results; poor discrimination; reference standards required for electronic testing.
Electrical resistance probe	Measure of moisture migration patterns in masonry walls.	Electrical resistance between two probes inserted into the test component is measured. The resistance decreases with increased moisture content.	Equipment is inexpensive, simple to operate, and many measurements can be made rapidly.	Not reliable at high moisture content; precise results are not usually obtained.
Flat jack	Determination of in situ stresses and deformability.	A portion of a mortar joint is removed and a flat jack installed in the slot. Deformations are measured before the slot is cut, and the jack pressure required to restore deformation is recorded.	Can determine stress in the wall in cases where loadings or displacements are unknown or difficult to quantify. Hi13h reliability.	Numerous measurements are required to obtain a full picture of deformation and stress. Usually, only one face of the masonry wall is accessible.

(Continued)

Table 6-22. (*Continued*)

Property	Purpose of test	Principle of operations	Advantages	Limitations
Fiber optics (probe holes with fiberscope)	Check the condition of materials in the cavity, such as thermal insulation, pipes, and electrical wiring installed in wall cavities; check for unfilled cores in reinforced masonry construction; check for voids along grouted stressed tendons.	A bundle of flexible, optical fibers with lens and illuminating systems is inserted into small boreholes, thus enabling a view of the cavities.	Direct visual inspection of otherwise inaccessible parts is possible.	Usually, probe holes must be drilled; probe holes must connect to a cavity.
Gamma radiography	Locates internal cracks, voids, and variations in the density and composition of materials. Locates internal parts in a structural component (e.g., reinforcing steel in masonry).	Gamma radiation attenuates when passing through a structural component. Extent of attenuation controlled by the density and thickness of materials of the structural component. A photographic film record is usually made and analyzed.	Portable and relatively inexpensive compared to x-ray radiography, internal defects can be detected, applicable to a variety of materials.	Radiation intensity cannot be adjusted; long exposure times may be required; dangerous radiation; two opposite surfaces of the component must be accessible. A special license is required to obtain gamma isotope sources.

(*Continued*)

Table 6-22. (*Continued*)

Property	Purpose of test	Principle of operations	Advantages	Limitations
Impact echo	Thickness and integrity of concrete and masonry.	Impact the test surface and use a displacement or accelerometer transducer to measure resonant echo responses from thickness and/or discontinuities such as cracks and voids	A one-sided access test Can detect unfilled collar joints as well as cracks, voids, and thicknesses in masonry	Cannot penetrate open cracks or voids.
Infrared thermography	Detection of delaminations of masonry, water intrusion in spray rack testing	An infrared camera records surface temperatures to reveal delaminations, water intrusions, and other discontinuities in masonry	Easy to use with one-sided access and remote access, as long as the structure is in view	Requires heat source, typically solar, which can be problematic Stains and discolorations influence results

(*Continued*)

Table 6-22. (*Continued*)

Property	Purpose of test	Principle of operations	Advantages	Limitations
Physical measurements	Physical documentation of observed conditions.	Physical measurement of alignment, crack size and location, differential settlement of structure, deflections, displacement, and distortion by means of surveying tools and instruments.	Easy to perform. No special equipment. Low cost.	Physical access required. This may require scaffolding and could be costly.
Ultrasonic pulse velocity	Internal discontinuities can be located and their size estimated.	Operates on the principle that vibrational wave propagation is affected by the quality of masonry; pulse waves are induced in materials, and those reflected back are detected; both the transmitting and the receiving transducers are usually contained in the same probe.	Excellent for rapid surveys of large areas.	Good coupling between transducer and test substrates is critical; interpretation of results can be difficult; the density and amount of aggregate may affect results; calibration standards required.

(*Continued*)

Table 6-22. (*Continued*)

Property	Purpose of test	Principle of operations	Advantages	Limitations
Visual examination	Evaluation of the masonry surface condition; determining deficiencies in joints; determining differential movement in joints; determining warping, bulging, or sagging of components. Visual examination includes sampling of units.	Visual examination with or without optical aids, photographic records, or other low-cost tools.	Usually low cost; rapid results except for the surveying method.	Trained evaluator required; primary evaluation confined to the surface of the structure.
Water penetration	Determine resistance to leakage and water penetration. The test is basically a laboratory test.	The test chamber placed against masonry with controlled temperature and humidity conditions. Water is applied with increased air pressure, and prime dampness is observed.	Provides useful information on the quality of material coatings, leakage, and workmanship. ASTM E514 intended for laboratory use. ASTM C1602 intended for field use.	Expensive. The opposite side of the wall may not be accessible to observe dampness.

Table 6-23. Evaluation of Properties of Masonry Units.

References: noted tables in Section 6.1.13.2 and 6.3.13.2	Table 6-21	Table 6-21	Table 6-21	Table 6-22	Table 6-21	Table 6-21	Table 6-21	Table 6-22
Evaluation procedure / Physical property	Compressive tests	Efflorescence test	Freeze–thaw resistance test	Measurement tests	Modulus tests	Moisture-related tests	Properties of finish tests	Visual examination
Absorption						•		
Adhesion	•							
Area in cored units				•				
Autoclave crazing							•	
Bulk-specific gravity						•		
Chemical resistance							•	
Compressive load	•							
Compressive strength	•							
Dimensions				•				
Distortion				•				
Drying shrinkage						•		
Efflorescence		•						
Freeze–thaw resistance			•					
Imperviousness						•		
Initial rate of absorption						•		

Table 6-23. (*Continued*)

References: noted tables in Section 6.1.13.2 and 6.3.13.2	Table 6-21	Table 6-21	Table 6-21	Table 6-22	Table 6-21	Table 6-21	Table 6-21	Table 6-22
Physical property \ Evaluation procedure	Compressive tests	Efflorescence test	Freeze–thaw resistance test	Measurement tests	Modulus tests	Moisture-related tests	Properties of finish tests	Visual examination
Modulus of elasticity					•			
Modulus of rupture					•			
Moisture content						•		
Moisture expansion						•		
Net area				•				
Opacity								•
Saturation coefficient							•	
Secant modulus of elasticity						•		
Size				•				
Solubility								•
Splitting tensile strength	•							
Surface Texture								•
Thermal expansion								•
Warpage				•				
Weight determination							•	
Young's modulus					•			

Table 6-24. Evaluation of Properties of Masonry Assemblages

References: See Table 6-21	Table 6-21	Table 6-21	Table 6-21	Table 6-21
Physical property \ Evaluation procedure	Vertical load tests	Transverse load tests	Shear tests	Deformability tests
Compressive strength	•			
Concentrated load capacity	•			
Deformability				•
Flexural tensile strength		•		
Modulus of rupture		•		
Secant modulus of elasticity		•		
Shear strength			•	
Tensile load		•		
Transverse load		•		

Table 6-25. Evaluation of Properties of Mortar and Grout.

References: See Table 6-21	Table 6-21
Physical property \ Evaluation procedure	Petrographic tests
Aggregate characteristics	•
Air content	•
Mortar and grout characteristics	•
Paste characteristics	•

Table 6-26. Evaluation of Physical Condition of Masonry Units and Masonry Assemblages (Environmental).

References: See Sections 6.1.13.2 and 6.3.13.2	Table 6-22	Table 6-3	Table 6-3	Table 6-22	Table 6-22	Table 6-22	Table 6-22	Table 6-22	Table 6-22	Table 6-22
Evaluation procedure / Physical condition	Acoustic impact	Electrical resistance probe	Fiber optics	Gamma radiography	Physical measurements	Ultrasonic pulse velocity	Visual examination	Water penetration	Impact echo	Infrared thermography
Air leakage					•		•			•
Blistering							•			•
Calcite streak							•			
Corrosion		•	•				•		•	
Coving							•			
Cracking	•	•	•	•	•	•	•		•	•
Crazing							•			
Crumbling							•			
Cryptoflorescence			•				•			

(Continued)

Table 6-26. *(Continued)*

References: See Sections 6.1.13.2 and 6.3.13.2	Table 6-22	Table 6-3	Table 6-3	Table 6-22	Table 6-22	Table 6-22	Table 6-22	Table 6-22	Table 6-22	Table 6-22
Evaluation procedure / Physical condition	Acoustic impact	Electrical resistance probe	Fiber optics	Gamma radiography	Physical measurements	Ultrasonic pulse velocity	Visual examination	Water penetration	Impact echo	Infrared thermography
Delamination	•			•		•	•		•	•
Deterioration		•	•				•			
Dew point					•					
Driving rain index					•					
Efflorescence							•			
Erosion							•			
Exfoliation							•			
Flaking							•			

(Continued)

Table 6-26. (*Continued*)

References: See Sections 6.1.13.2 and 6.3.13.2	Table 6-22	Table 6-3	Table 6-3	Table 6-22	Table 6-22	Table 6-22	Table 6-22	Table 6-22	Table 6-22	Table 6-22
Evaluation procedure / Physical condition	Acoustic impact	Electrical resistance probe	Fiber optics	Gamma radiography	Physical measurements	Ultrasonic pulse velocity	Visual examination	Water penetration	Impact echo	Infrared thermography
Friability							•			
Interstitial condensation		•	•				•			
Lime run							•			
Peeling							•			
Pitting							•			
Rising damp		•	•		•		•			
Salt fretting							•			
Spalling							•			

(*Continued*)

Table 6-26. (*Continued*)

References: See Sections 6.1.13.2 and 6.3.13.2	Table 6-22	Table 6-3	Table 6-3	Table 6-22	Table 6-22	Table 6-22	Table 6-22	Table 6-22	Table 6-22	Table 6-22
Evaluation procedure / Physical condition	Acoustic impact	Electrical resistance probe	Fiber optics	Gamma radiography	Physical measurements	Ultrasonic pulse velocity	Visual examination	Water penetration	Impact echo	Infrared thermography
Staining							•			
Sugaring							•			
Surface condensation							•			
Surface deposits							•			
Surface induration							•			
Surface temperature							•			
Tidemark							•			
Transition area			•				•			
Warpage							•			
Water penetration/ permeance		•					•	•		•
Weathering							•			

Table 6-27. Evaluation of Physical Conditions of Masonry Units and Masonry Assemblages (Structural).

References: See Table 6-22	Table 6-22	Table 6-22
Evaluation procedure / Physical condition	Physical measurements	Visual examination
Bowing	•	•
Bulging	•	•
Chipping		•
Condition of lintels (shelf angles)		•
Control joints	•	•
Cracking	•	•
Deflection	•	•
Detachment		•
Displacement	•	•
Distortion	•	•
Lintels		•
Plumbness	•	•
Sagging	•	•
Settlement	•	•
Spalling		•
Ties		•
Warping	•	•

Table 6-28. Evaluation of Physical Conditions of Masonry Units and Masonry Assemblages (Architectural).

References: See Table 6-22	Table 6-22	Table 6-22
Evaluation procedure / Physical condition	Physical measurements	Visual examination
Condition of flashing	•	•
Condition of roofing		•
Control joints	•	•
Damp course		•
Dampproofing		•
Expansion joints		•
Gutters and Downspouts		•
Location		•

Table 6-29. Evaluation of Physical Conditions of Mortar and Grout.

References: See Tables in Section 6.1.13.2	Table 6-3	Table 6-3	Table 6-3	Table 6-3	Table 6-3
Evaluation procedure / Physical condition	Cover meter	Fiber optics	Gamma radiography	Physical measurement	Visual examination
Air content				•	
Alignment				•	•
Bed joint levelness				•	•
Bed width variation				•	•
Condition					•
Cracking		•	•	•	•
Exposure of joint Reinforcement	•	•	•		
Friability					•
Hardness					•
Pitting					•
Sandiness					•
Voids					•

6.4 CONDITION ASSESSMENT OF WOOD

The condition of a wood structure is best determined through understanding its material properties and physical conditions by visual examination in conjunction with applicable testing. Member species and physical and environmental conditions are discussed in Section 6.4.1. Test methods that can be used to evaluate the wood members are provided in Section 6.4.13. A condition assessment of wood structures may require determinations of species, quality of wood, and environmental conditions.

Proper assessment and subsequent evaluation should be utilized to reveal the wood's ability to sustain loads and endure impacts of the environmental conditions to which it is being or will be subjected.

Basic information for conducting an engineering analysis of existing wood structures requires knowledge about how new structures are designed. The *AWC National Design Specification (NDS) for Wood Construction* (AWC 2018) should be referenced for this purpose. Older, existing structures may require reference to the earlier editions of the NDS and the *AITC Timber Construction Manual* (AITC 2012) for previously used design methods and requirements. Information for conducting an engineering evaluation can be found in *TFEC Guide to Structural Evaluation of Existing Timber Structures and Structural Timber Assessment Using Nondestructive Evaluation* (2019a).

6.4.1 Reasons for Conducting a Wood Condition Assessment

Although a structural condition assessment can be conducted on a wood structure for a number of reasons when considering strengthening, repairs, or rehabilitation, the most common *technical* reasons often come under the following categories:

Deterioration Concerns: There is a need to determine whether biological deterioration may be present, to locate and quantify the extent of deterioration, and assess its impact on material properties.

Strength or Stiffness Questions: There is a need to determine the capacity and grade of individual wood members or components.

Connections: There is a need to verify the presence of connectors; determine whether metal, wood, adhesive, or other fastener types that form joints between wood and/or other types of structural members are deteriorated; or assess the capacity of the connection.

Fire: There is a need to understand the effects of fire or elevated temperature on wood members. Fire affects wood construction through combustion and elevated temperature. Wood members may lose section because of combustion but may also lose strength because of elevated temperatures over a sustained period. In addition, connections may be weakened because of differential thermal expansion relative to the wood, resulting in loose connections.

Prior or Proposed Modifications and Alterations: Observation of modifications and alterations to the original building may warrant additional consideration when conducting a wood condition assessment in those areas to ensure that unintended consequences have not compromised the wood.

Note that over time, there has been a greater emphasis on energy efficiency of buildings. For existing buildings, addressing energy efficiency is beyond the scope of a typical wood condition assessment; however, modifications to existing buildings to make them more energy-efficient, such as adding insulation in exterior walls or at roofs, can alter the moisture and temperature environment sufficiently that condensation, wood decay, or other detrimental conditions may develop.

6.4.2 Wood as a Material

Wood is a biological material. Wood products can be made of solid wood, such as lumber, timbers, poles, or piles, or manufactured by reconstituting wood into engineered wood products such as panel products, glued-laminated timber, I-joists, structural composites, cross-laminated timber, and many other products.

Because it is a biological material, wood has natural growth characteristics such as knots, slope of grain, and seasoning checks that develop as the wood dries, and that can be redistributed in engineered wood products to reduce the inherent variability of the physical and mechanical properties of solid wood. As an engineering material, wood, particularly when used as solid sawn and some wood-based panel products, is considered to be orthotropic and nonhomogeneous, with mechanical properties dependent on fiber orientation in a product. The term *fiber* is used synonymously with *grain,* meaning the direction corresponding to the long axis of the wood cells. Because of the orthotropic nature of wood, there can be considerable variability in the physical and mechanical properties of wood products.

Understanding the inherent variability of the wood properties of either solid wood products or engineered wood products is essential when conducting a condition assessment to assess wood behavior and establish material properties. In addition to the inherent variability of wood properties, common terminology is critical for communicating the findings of a structural condition assessment. Too often, terms are reported that are confusing because of the lack of use of standard terminology. A condition assessment is intended to provide information needed to determine the adequacy of a structure to perform its intended function and provide the decision-maker with the ability to make more informed decisions on the basis of known material properties and conditions.

6.4.3 Terminology Related to Wood Characteristics and Properties

Regardless of whether a wood product is solid sawn lumber or timber or one of the engineered wood products, the basic terminology that is relevant to condition assessments is the same (ASTM D9, "*Standard Terminology Relating to Wood and Wood-Based Products*) (ASTM 2020).

Throughout the text, references are given for specific publications referenced in the body of the text. Those references are listed in the

Appendixes B1 through B4, as are sources of additional information that may not be specifically referenced within the body of the text.

The following list from ASTM D9 gives common terms that are relevant to wood when conducting a structural condition assessment.

Wood Species: Woods species are classified into two broad groups, often termed "softwoods" and "hardwoods," but more accurately as "coniferous species" and "deciduous species" (USDA 2011). The hardwood/softwood terminology has no reference to the actual hardness of the wood. The classification of wood is not as important as the specific wood species. Wood is sorted by species for listing physical and mechanical properties. It is important to note that physical and mechanical properties found in various tables are typically species average values and are not the actual value for a particular wood product. Most often, a mean value is published, and occasionally, the variation may be listed as either the standard deviation or the coefficient of variation. Physical and mechanical properties listed in technical publications, such as the *Wood Handbook: Wood as an Engineering Material* (Ross 2021), are useful in understanding why wood products are, or are not, performing in a particular building but should not be confused with referenced design values published in standards such as the *National Design Specification for Wood Construction*. Such standards typically have values that are calculated from species average values and standard deviation (to determine the lower 5th percentile value to account for the inherent variability of wood properties), plus a safety factor to arrive at a reference design value.

- Softwoods (coniferous species) are usually cone-bearing trees with needle-like leaves. The term *softwood* has no reference to the actual hardness of the wood. The majority of wood structural members are manufactured from softwoods. Examples include Douglas Fir, pines, spruce, fir, and cedar. Some species, such as Southern Pine, consist of several species that are grouped together for marketing and design purposes. For example, existing buildings may have been built with longleaf pine, but no values are published today for longleaf pine; it is grouped with related pines and assigned design values as the group Southern Pine.
- Hardwoods (deciduous species) typically have broad leaves rather than needles, although not always. The term hardwood has no reference to the actual hardness of the wood. These woods are, in general, used architecturally for flooring, trim, and paneling. However, they are also used as load-carrying members, most often in timber-frame structures. Many older existing (historic) structures in the eastern United States were built of oak or chestnut

framing lumber. Examples of commonly used species today include oak, ash, cherry, and walnut. Most, but not all, tropical species used in construction are considered hardwoods.

Allowable Properties: Allowable property is typically referred to as the Reference Design Value and is the mechanical property of a material as specified by the National Design Specification for Wood Construction for design use.

American Softwood Lumber Standard: Voluntary product standard of the US Department of Commerce that serves as a basis for the preparation and revision of industry-sponsored grading rules for lumber.

Annual Ring: Growth layer produced by the tree in a single growth year, including earlywood and latewood.

Basic Stress *(archaic)*: Term once used for an allowable property for clear, straight-grained lumber. Often found in older building codes, standards, and technical references.

Brashness: Condition that causes some pieces of wood to be relatively low in shock resistance for the species and, when broken in bending, to fail abruptly without splintering at comparatively small deflections.

Characteristic Value: Structural property estimate, typically a population mean, median, or tolerance limit value derived from test data that is representative of the population being sampled. Test data are typically adjusted to standardized conditions of temperature, moisture content, and characteristic size.

Check: Separation of the wood along the fiber direction that usually extends across the rings of annual growth, commonly resulting from stresses set up in wood during seasoning.

- *End check*: A seasoning check occurring on the end of a board or other pieces of wood.
- *Heart check:* A check that extends across the growth layers in one or more directions from the pith toward, but not to, the surface of a piece of wood. A synonym is *pith check*.
- *Surface check*: A check occurring on the surface of a piece of wood, usually on the tangential face not extending through the piece.

Clear: Free of defects or imperfections.

Compression Wood: Abnormal wood formed typically on the lower side of branches and inclined trunks of softwood trees. Compression wood is identified by its relatively wide annual rings, usually eccentric, relatively large amount of latewood, sometimes more than 50% of the width of the annual rings in which it occurs, and its lack of demarcation between earlywood and latewood in the same annual rings. Compression wood shrinks excessively lengthwise, compared to normal wood.

Cross Break: Separation of the wood cells across the grain. Such breaks may be caused by internal stress resulting from unequal longitudinal shrinkage or external forces.

Decay: Decomposition of wood substance caused by the action of wood-destroying fungi, resulting in softening, loss of strength and weight, and often in change of texture and color.

Advanced (or typical) decay: Older stage of decay in which the destruction is readily recognized because the wood has become punky, soft and spongy, stringy, ringshaked, pitted, or crumbly. Decided discoloration or bleaching of the rotted wood is often apparent.

Incipient decay: Early stage of decay that has not proceeded far enough to soften or otherwise perceptibly impair the hardness of the wood.

Defect: Any irregularity or imperfection in a tree, log, piece, product, or lumber that reduces the volume of sound wood or lowers its durability, strength, or utility value.

Dense: (1) Term used in stress grading of certain softwood species to signify a high specific gravity. (2) a visual estimate of high specific gravity.

Density: Oven-dry mass of solid wood is termed "density", which is used to establish the specific gravity of wood. Some physical and mechanical properties are correlated with density. Density is most often used for determining the capacity of connections using wood members. In general, higher densities result in a higher capacity of connections. Density is also used with some nondestructive testing techniques, but caution should be exercised because density is correlated with mechanical properties, typically modulus of elasticity and bending strength, but it does not account for the effect of natural growth characteristics that may be present, such as knots. Engineered wood products reduce the likelihood of natural growth characteristics affecting the correlation between density and mechanical properties by eliminating or redistributing the characteristics throughout the product.

Dimensional Stabilization: Treatment of wood to reduce swelling and shrinking caused by changes in its moisture content.

Durability: General term for permanence or resistance to deterioration. Frequently used to refer to the degree of resistance of a species of wood to attack by wood-destroying fungi under conditions that favor such attack. In this connection, the term "decay resistance" is more specific.

Duration of Load: Duration of stress or the time during which a load acts on a member. In wood, a design consideration for modifying allowable stresses, based on the accumulated loadings anticipated in the life of a structure.

E-Rated Lumber: Lumber graded for use in manufacturing glued-laminated timber by nondestructive measurement of a modulus of elasticity (*E*) and by visual inspection in accordance with the grading rules of the applicable grading or inspection agency.

Earlywood: Less dense, large-celled part of the growth layer formed first during the annual growth cycle. A synonym for *Springwood*.

Equilibrium Moisture Content: Moisture content at which wood neither gains nor loses moisture to the surrounding air.

Fiber Saturation Point: Moisture content at which the cell walls are saturated with water (bound water), and no water is held in the cell cavities by capillary forces. It is usually taken as 25% to 30% moisture content on the basis of weight when oven dry.

Fire Retardant: Having or providing comparatively low flammability or flame-spread properties.

Form Factor: Factor used in the calculation of the bending strength of wood members to compensate for a nonrectangular shape. Defined as the ratio of the modulus of rupture of a member of desired shape to that of a "standard" member.

Fuzzy Wood: Damage caused by salt uptake into the wood, also referred to as *salt kill* or *salt defibration*. It is sometimes seen in wood that is chronically exposed to salt, such as marine pilings, bridge decks where salt is used as a deicer, utility poles that are splashed with road salt, or wood associated with salt storage.

Grade: Designation of the quality of logs or of a manufactured piece of wood. Visual grading is the oldest stress-grading method for lumber and timbers. It is based on the premise that the mechanical properties of lumber differ from the mechanical properties of clear wood on the basis of the many growth characteristics that can be seen and judged visually. These growth characteristics are used to sort lumber into stress grades. To be graded and marked with a grade stamp, all six surfaces of a member must be examined for conformance with explicit grading rules. Although only a certified lumber grader may stamp lumber with a grade stamp, anyone with a sufficient understanding of the grading rules can grade lumber and timber, but they cannot apply a stamp. Certified lumber graders are typically certified to grade particular products (e.g., lumber but not timber), and the grade stamp is associated with a specific mill for those products. A grade stamp on a member provides information on which to base its appropriate use. Without a grade stamp, someone trained in grading may be needed to assign the appropriate grade.

Machine stress rating is based on an empirical relationship between the modulus of elasticity and the bending strength of lumber. Measurement of the modulus of elasticity is the sorting criteria used in this method of grading. Machines operating at high rates of speed measure the modulus of elasticity of individual pieces of lumber and sort according to a bending strength/stiffness grade class. For in situ applications, measurement of stress wave velocity and an estimate of density have been used to estimate the modulus of elasticity. Using the established empirical relationships between the modulus of elasticity and the modulus of rupture, member strength or structural grade can be estimated for specific wood species.

Grain: Direction, size, arrangement, appearance, or quality of the fibers in lumber or other wood products. To have a specific meaning, the term must be qualified (see ASTM D9 for additional discussion).

- *Cross grain*: Fiber deviation from a line parallel to the sides of the piece. Cross grain may be either diagonal or spiral grain or a combination of the two.
- *End grain*: The grain pattern exposed when ends of logs or timbers, dimension, boards, and other wood products are cut perpendicular to the fiber direction.
- *Flat grain*: The grain pattern resulting when lumber has been sawed in a plane approximately perpendicular to the radius of the log. Lumber is considered flat grained when the annual growth rings make an angle of less than 45 degrees with the surface of the piece.
- *Plainsawn*: Synonym for *flat grain*.
- *Quartersawn*: Synonym for *edge grain*.
- *Raised grain*: A condition of the surface of dressed lumber in which the hard latewood is raised above the softer earlywood but not torn loose from it.
- *Rift sawn*: Synonym for *edge grain*.
- *Slash grain*: Synonym for *flat grain*.
- *Spiral grain*: Wood in which the fibers take a spiral course about the trunk of a tree instead of the normal vertical course. The spiral may extend in a right-handed or left-handed direction around the tree trunk. Spiral grain is a form of cross grain.
- *Straight-grained wood*: Wood in which the fibers run parallel to the axis of a piece.
- *Vertical grain*: A synonym for *edge grain*.

Hardness: Term relating to the capacity of wood to withstand denting and abrasion; for purposes of comparison, hardness in wood is often measured as the load in pounds required to embed a 0.444 in. ball to one-half its diameter (Methods D143).

Heartwood: Inner layer of a woody stem wholly composed of nonliving cells and usually differentiated from the outer enveloping layer (sapwood) by its darker color. It is usually more decay-resistant than sapwood.

Holes: Openings in or through lumber that may extend partially or entirely through a piece and may be from any cause.

Incise: To make slit-like lacerations typically parallel to the grain in the lateral surface of timbers that are resistant to treatment, so that a deeper and more uniform penetration of preservative may be obtained.

Joint: Junction of two or more pieces of wood. Usually implies a structural junction in which stress is transmitted from one piece to another. Also called a *connection*.

Juvenile Wood: Wood formed adjacent to the pith, characterized by progressive change in cell dimension, different microstructure from mature wood, and greater shrinkage parallel to the grain. In softwoods, juvenile wood is characterized by lower specific gravity and strength.

Knot: That portion of a branch or limb that has been surrounded by subsequent growth of the wood of the tree. As a knot appears on the cut surface, it is merely a section of the entire knot, its shape depending on the direction of the cut.

- *Edge knot*: (1) Knot located at the edge of the face in a piece of lumber. (2) in stress grading under Practice D 245, a knot whose perimeter falls within one-sixth of the knot width from the edge of the piece.
- *Spike knot*: Knot cut from 0 to 45 degrees to the long axis of the limb.

Lamination: Full-width and full-length layer contained in an assembly or product bonded together with adhesive or mechanical fasteners. The layer itself may be composed of one or several wood pieces in width or length.

Latewood: Denser, smaller-celled, later-formed part of a growth layer. Synonym for *summerwood*.

Longitudinal: In general, the direction parallel to the grain of wood. Sometimes the long axis of a wood specimen.

Lower Tail: Portion of an ordered data set consisting of all test specimens with the lowest property values (e.g., lowest strengths).

Moisture Content: Amount of water contained in the wood, usually expressed as a percentage of the mass of the oven-dry wood. (1) freshly sawed wood, or wood that has received no drying is referred to as being **green** or unseasoned; also, lumber that may have become wet to above the fiber saturation point may be referred to as being in the "green condition"; and (2) wood above a stipulated moisture content, as lumber above 19% moisture content in accordance with the American Softwood Lumber Standard.

Moisture Gradient: Condition of graduated moisture content between successive thickness zones of wood that may be losing or absorbing moisture.

Nominal Size: As applied to products such as lumber, traditionally the approximate rough-sawn commercial size by which it is known and sold in the market. Actual rough-sawn sizes may vary from the nominal. Reference to standards or grade rules is required to determine nominal/actual finished size relationships.

Pitch: Term applied to the resin occurring in the wood of certain conifers.

Pith: Small, soft tissue occurring in the structural center of a tree trunk, branch, twig, or log.

Preservative: Chemical substance which, when suitably applied to wood, makes the wood resistant to attack by fungi, insects, marine borers, or weather conditions.

Pressure-Treated Wood: Wood treated by applying pressure to force the preservative into it.

Punky: Wood that has intermediate decay. The wood is soft and easy to penetrate but has not yet deteriorated to the point where there is a void.

Radial: Coincident with the radius from the pith to the circumference of the tree or log.

Rate of Growth: Rate at which a tree has grown. The unit of measure is the number of annual growth rings per inch, measured radially in the trunk or lumber cut from the trunk.

Sawn Lumber: Product of a sawmill that is produced by cutting a long or cant lengthwise to produce square or rectangular members.

Sapwood: Wood containing some living cells and forming the initial wood layer beneath the bark of the log. The thickness of the sapwood layer varies by species and may be lighter in color than heartwood. Under most conditions, the sapwood is more susceptible to decay than heartwood.

Seasoning: Drying, the term often applied to the process of removing moisture from wood to achieve a moisture content appropriate for the performance expected of the final product.

- *Air-Dried*: Dried by exposure to air, usually in a yard, without artificial heat.
- *Dry*: Seasoned; in softwood lumber, the abbreviation S-Dry means not in excess of 19% moisture content at time of surfacing, in accordance with recognized standards.
- *Kiln-Dried*: Dried in a closed chamber in which the temperature and relative humidity of the circulated air can be controlled.
- *Oven Dry:* Dried in an oven to remove all moisture. The temperature employed usually is 101 to 105 °C (214 to 221°F) in accordance with ASTM D2016 *Methods of Test for Moisture Content of Wood* (withdrawn and superseded by ASTM D4442) (ASTM 2020).

Shake: Longitudinal separation of the wood. Usually referred to as *ring shake*—a shake occurring in the plane of the growth rings in the outer portion of the latewood for partial or entire encirclement of the pith, occasionally moving radially to an adjacent latewood ring.

Shrinkage: Reduction in dimensions caused by lowering the moisture content below the fiber saturation point.

Size Factor: Factor included in deriving the allowable bending stress for rectangular members, which takes into account the somewhat lower unit strength developed in larger members compared to smaller members.

Slope of Grain: Fiber deviation from a line parallel to the sides of the piece; also called *cross grain*, it is measured as the deviation of the fibers from being parallel to the edge of the piece as slope (rise over run), such as 1 to 10, or 1 in. of deviation from parallel in 10 in. of length along the piece. Cross grain may be either diagonal or spiral grain or a combination of the two.

Sound Wood: Wood free of any form of decay, incipient or advanced, and from insect holes.

Specific Gravity: As applied to wood, the ratio of the oven-dry weight of a sample to the weight of a volume of water equal to the volume of the sample at some specific moisture content, as green, air dry, or oven dry.

Split: Separation of the wood parallel to the fiber direction caused by the tearing apart of the wood cells.

Stain: Discoloration in wood that may be caused by such diverse agencies as microorganisms, metal, or chemicals.

Strength: (1) Ability of a member to sustain stress without failure. (2) in a specific mode of test, the maximum stress sustained by a member loaded to failure.

Strength Ratio: Hypothetical ratio of the strength of a member to the strength it would have if no weakening defects were present.

Strength-Reducing Defects: Imperfections affecting strength, such as checks, compression wood, cross grain, decay, knots, shakes, and splits.

Tangential: Strictly, coincident with a tangent at the circumference of a tree or log, or parallel to such a tangent. In practice, however, it often means roughly coincident with a growth ring.

Tension Wood: Abnormal form of wood found in the upper side of the bole and branches of leaning trees of some hardwood species and characterized by the presence of gelatinous fibers and excessive longitudinal shrinkage.

Thickness: Lesser dimension perpendicular to the long axis of lumber or timber.

Transverse: Directions in wood at right angles to the wood fibers. Includes radial and tangential directions. A transverse section is a section through a tree or timber at right angles to the pitch.

Treatment: Act or manner of treating wood; the quantity of preservative or other substances specified or used to treat wood. Also refers to *remedial treatment*, which is typically the in situ application of a preservative to retard decay or insect activity in wood that is in a structure.

Weathering: Mechanical or chemical disintegration and discoloration of the surface of wood that is caused by exposure to light, the action of dust and sand carried by winds, and the alternate shrinking and swelling of the surface fibers with the continual variation in moisture content brought by changes in the weather. Weathering is not decay.

Width: Greater dimension perpendicular to the long axis of lumber or timber.

Wood: Tissues of the stem, branches, and roots of a woody plant lying between the pith and the cambium, serving for water conduction, mechanical strength, and food storage, and characterized by the presence of tracheids or vessels.

Wood-Based Product: Manufactured product that has substantial wood content and measurable properties determined in accordance with a consensus standard.

Wood-Destroying Organisms: Principally the decay-producing fungi, beetles, termites, carpenter ants, marine borers, and others.

Wood Substance: Moisture-free material of which wood is composed.

6.4.4 Physical and Mechanical Properties of Wood

Wood is orthotropic. In other words, it has different properties in the direction of the three mutually perpendicular axes—longitudinal, radial, and tangential. Traditionally, the fundamental mechanical properties of solid wood used to establish referenced design values were determined from tests performed on small, clear samples. Design values for solid structural wood members are lower than published clear wood properties to account for the presence of defects (i.e., the structural grade) and include a reduction factor for material variability and end-use condition (type of loading, wet or dry service condition). The published design values are based on test data and procedures published by ASTM that demonstrate the engineering performance of the material. Solid wood products are graded in accordance with procedures promulgated by one of several forest product industry associations. The design values given in the building code for solid wood products are established by the American Wood Council and published as the *National Design Specification for Wood Construction* (AWS 2018). Design values for engineered wood products are published by trade organizations, the International Code Council as ICC-ES Evaluation Reports, or as proprietary data by manufacturers.

Bending: Allowable bending strength is based on a percentage of the ultimate load-carrying capacity as indicated by a modulus of rupture.

Compression Parallel to Grain: Measure of wood's resistance to axial stresses along the axis of the fibers. These stresses are important in evaluating compression members and bearing capacity of the fasteners.

Compression Perpendicular to Grain: Measure of wood's resistance to load when applied perpendicular to the axis of the fibers. An example is a beam resting on a post. The length and location of the bearing area are factors in determining the allowable stress.

Fatigue Strength: When the repetitions of design stress are expected to be more than 100,000 cycles during the normal design life, an allowable design stress less than the allowable static design stress should be considered.

Modulus of Elasticity: Modulus of elasticity of wood is dependent on the orthotropic direction under consideration. However, design moduli of elasticity are based on bending tests of lumber and are usually given only for the longitudinal direction. The modulus of elasticity for wood is not adjusted for the duration of load.

Shear Parallel to Grain: Horizontal shear is developed from parallel-to-the-grain bending stresses. Because of the structure of wood, shear resistance along the fibers controls shear capacity. Specific concerns include notches and holes in beams. Wood does not have vertical shear as a strength property.

Specific Gravity: Measure of density. For solid wood, the ratio of the oven-dry weight of a wood sample to the weight of a volume of water equal to the volume of the sample at some specific moisture content, as green, air-dry, or oven-dry. Specific gravity is used when determining the capacity of connections.

Tensile Strength Parallel to Grain: Clear wood intrinsically has high tensile strength parallel to the wood fibers. However, growth and manufacturing defects may markedly reduce a member's tensile strength when compared to clear wood strength.

Tensile Strength Perpendicular to Grain: Measure of the resistance of wood to forces acting across the wood fibers that tend to cause splits in a member. This is typically the weakest mechanical property for wood and, when exceeded, results in brittle fracture. Many structural failures can be traced to excessive tensile stresses perpendicular to the grain.

6.4.5 Wood Products

Solid Sawn Lumber and Timber: Dimensional lumber and timbers are the most widely used structural wood products in existing buildings. Over time, the ratio of solid sawn wood products to engineered wood products and wood composite panels will decrease, as these products either replace or augment solid sawn products.

- Dimensional lumber is defined as less than or equal to 4 in. in thickness, and timbers are defined as greater than or equal to 5 in. in thickness. The definitions are important when assigning structural grades and allowable design values. Lumber is graded under the National Grading Rule published within the American Softwood Lumber Standard PS-20 (US Department of Commerce). There is a single set of names and descriptions for lumber graded under these rules. Structural grades are assigned to specific species and size combinations for individual members by visual grading procedures (e.g., Select Structural, No. 1, No. 2) and/or by machine grading of the lumber nondestructively based on correlations between material strength (typically bending strength, in other words modulus of rupture) and stiffness (modulus of elasticity) [e.g., machine stress–rated grades 1.650 °F (898.8 °C), 2,100 °F (2,248.8 °C), 2,400 °F (1,315.5 °C)]. These stress grades are typically indicated by a grade stamp on the member. Lumber grade names and classifications are given in various grading agency specifications.
- Timbers are visually graded in accordance with procedures defined by the various grading agencies. Although no grading agency has responsibility for all commercially available species, coordination among agencies limits discrepancies in the grading rules. Most notably, timbers made of Southern Pine are called *timbers*, but timbers of all other species are segregated into either beams and stringers or posts and timbers on the basis of the difference between the thickness and the width of the timber. If the thickness and width of a non-Southern Pine timber are greater than 2 in. (e.g., a nominal 6 × 12), it is classified as a beam and stringer. If the thickness and width of a non-Southern Pine timber are no greater than 2 in. (e.g., a nominal 8 × 10), it is classified as a post and timber. All lumber and timber dimensions are given currently as nominal sizes with tolerances for actual dimensions defined in the *National Design Specification and Design Values for Wood Construction* (AWC 2018). However, in existing buildings, the actual dimensions are often different from today's nominal dimensions, and the actual size may need to be considered when determining either structural grade or allowable load.

Structural Composite Lumber, Including Laminated Products: Structural composite lumber, including laminated products, is made by bonding layers of lumber, veneer, and/or small wood particles (chips, flakes, etc.) together to form dimensioned lumber. If properly manufactured, such products have reduced material property variability compared to solid-sawn wood.

- *Structural Glued Laminated Timber (Glulam).* These members are made by face-gluing (as opposed to edge-gluing) laminations of solid sawn lumber to form the structural member. The grain of the lamination is parallel to the longitudinal axis of the member. Glulam can be architectural or structural and can be made to a range of widths, depths, and lengths. Grades and stress classifications of glulam timbers are described in *American National Standard for Wood Products–Structural Glued Laminated Timber*, A190.1 (ANSI/AITC 2022), and technical literature and data are provided by APA–The Engineered Wood Association and the American Institute of Timber Construction, including specifications, inspection, and repair procedures.
- *Laminated Veneer Lumber (LVL).* Most LVL products are proprietary and are made by gluing thin veneers with the grain of all veneers parallel to the longitudinal axis of the member. Uses for LVL include joists, beams, and truss chords. Grades and classifications for LVL are described by manufacturers' product information and ICC ESRs.
- *Cross-Laminated Timber (CLT).* CLT is a large-scale, prefabricated, solid engineered wood panel. It is manufactured with kiln-dried lumber boards or structural composite lumber laminated in alternating directions and bonded with structural adhesives to form a solid, straight rectangular panel. The panels are typically 20 in. or less thick, 2, 4, 8, or 10 ft in width, and up to 60 ft long. CLT is present within structures where high-load-bearing properties can be used, such as bridges, ancillary buildings, carports, and others. Lamination grades are either machine stress (four possible) or visually rated (three possible). Allowable bending capacities are available in the ANSI/APA PRG 320, *Standard for Performance Rated Cross-Laminated Timber* (APA 2019).
- *Products from Modified Wood.* Advances in technology have led to modified wood being used in construction applications. Acetylated wood and thermally modified wood are examples of new wood products available in the market. The technology for these processes was developed decades ago, but the commercial availability of the products is relatively recent, first in Europe and, more recently, in North America. No standards that govern these proprietary wood products are available yet,

and procedures or methodologies for conducting a condition assessment on these products are currently lacking. However, deterioration mechanisms are essentially the same as for other wood products, and the assessment of these products can use the same tools and methodologies. Although modified wood products are typically marketed as not losing strength over time, there is currently no means to assess their strength in situ or nondestructively.

Trusses: Trusses are assemblies that rely on the interaction of the various elements and their associated connectors. Several types of wood trusses exist. These include

- Heavy timber trusses connected with bolts, split rings, and heavy metal plates. The truss configurations can be infinite, but the configurations typically encountered during a condition survey may be parallel chord trusses at floor and roofs (top and bottom chords are parallel to each other) or bowstring trusses (the bottom chord is straight, whereas the top chord is in an arch shape). The truss spans vary from approximately 20 ft to more than 100 ft long. The top and bottom chords are double members, with the web members separating them. The bowstring curved top chords are typically glue-laminated members, except for older trusses, whereas the bottom chords are sawn lumber or timber members. The bottom chords typically have tension splice–bolted connections with a wood splice block between the members, steel plates on the exterior, and are typically located near the one-third points of the span. The ends of the truss contain a metal shoe to connect the top and bottom chord ends. The top chord bears on the vertical end plate of the shoe, whereas the bottom chord is connected with a tension-bolted splice. The trusses require cross-bracing members to make them stable. The design of heavy timber trusses is described in various references (Gurfinkel 1981, Hoyle and Woeste 1989, Breyer 2019, AITC 2012).
- Light wood trusses connected with metal plates or wood gussets. Light wood trusses are typically manufactured at a plant and shipped to the site and erected in one piece. They contain individual sawn lumber pieces that are connected with metal press plates at the chord–web and web–web member intersection. The trusses can be made in multiple configurations to meet the needs of the architect and engineer but are typically limited to the length and height that can be shipped to the job site. These trusses require either lateral bracing or bridging members to make them stable. Information on metal plate–connected wood trusses or metal webs can be found in individual manufacturers' product catalogs or from the Truss Plate Institute (TPI).

- Metal-webbed parallel chord trusses (Joists). Metal-webbed parallel chord trusses consist of either two engineered or sawn lumber top and bottom chords and metal tubular web members. The chord–web connections are typically single through-bolt connections. The trusses can be manufactured in numerous configurations, and to achieve longer spans (some more than 100 ft long), the trusses can be shipped in multiple sections and assembled at the site. These trusses require either lateral bracing or bridging members to make them stable.
- Timber Frames. Timber frames can be traditional construction, consisting of timbers connected by timber joinery, and may not use any metal fasteners. Post and timber construction also consists of timbers but, typically, uses bolts, nails, and metal plates to connect the timbers. Although the primary building units may be solid sawn timbers, glue-laminated timber, or structural composite lumber, the structure can present unique analytical challenges, particularly if tradition timber joinery is used. Caution should be exercised when assessing timber frame structures because they often behave as a system and the mortise-and-tenon joinery is not well suited to most structural analysis programs. As a result, conclusions regarding the adequacy of these systems are often overly conservative. An understanding of the joinery and connections is key to understanding the behavior of timber frames. The Timber Framers Guild provides documentation on the design, analysis, and construction of timber frames.

Structural Wood Panel Products

- *Plywood*. Plywood is made using a balanced construction of thin veneers glued with their grain alternating 90 degrees in adjacent layers. Plywood grades and stress classifications are described by the APA - *The Engineered Wood Association Engineered Wood Construction Guide E30* (APA 2020).
- *Wood Composite Panels*. Wood composite panels are manufactured by bonding veneer faces to reconstituted wood cores or bonding with adhesive small wood particles (chips, flakes, etc.) Examples include oriented strand board (OSB), medium-density fiberboard (MDF), hardboard, and particleboard. OSB is the only wood composite panel with structural properties intended for load-bearing applications. Grades and classifications for composite panels are described by the APA - *The Engineered Wood Association Engineered Wood Construction Guide E30* (APA 2020).

Stressed-Skin Panels and Structural Insulated Panels: Stressed-skin panels and structural insulated panels (SIPS) are constructed using wood panels on one or both faces and rigid insulation as the core. The face plies are typically OSB. Specifications for SIPS are given by the Structural Insulated Panel Association. In addition to ease of installation, SIPS are fabricated in a manufacturing facility under controlled conditions, thereby reducing the variability that would result from field-fabricated panels.

Combined Wood Structural Elements: These include highly efficient structural products manufactured by combining different wood elements. Examples include structural composite lumber, wood I-beams, builtup beams, and stress-skin panels. Product classifications are described by manufacturers' product information. Structural components combining wood and other materials are found in building construction, including timber arches, rigid frames tied together with steel rods, and flitch beams.

6.4.6 Mechanisms of Wood Deterioration

Wood is susceptible to attack by wood decay fungi, wood-destroying insects, and marine borers. Bacteria can also degrade wood but to a much lesser and slower degree than the other biological organisms. Deterioration of wood in buildings is most often the result of wood decay or insect attack but is influenced by moisture content, the duration of load and load history, mechanical damage, and fire.

Fungal Attack: Primary mechanism of deterioration of wood used in structures is attacked by decay fungi (Emmerich 1986). Decay fungi should not be confused with mildew, mold, or stain fungi. Mildew grows on the surface of wood (and many other materials) and does not degrade the wood. Mold and stain fungi penetrate the wood surface and may discolor the surface of the wood but also do not degrade the wood such that a loss of strength results. Decay is caused by specific types of fungi, which degrades the chemical constituents of the wood and results in a loss of strength. For decay fungi to be active, the wood must be available as food for the fungi; the moisture content of the wood must, typically, be above 20%; temperatures must be approximately 40 to 120 °F (4 to 49 °C), and oxygen is required for fungi growth. There are three stages of decay development: (1) incipient decay where the infected wood is subjected to decay fungi in long finger-like growths; incipient decay is difficult to identify visually; (2) intermediate decay where the wood becomes soft and punky with obvious loss of strength; and (3) advanced decay where no strength remains, and wood integrity is lost. Wood decay fungi reproduce through spores, which are, for all practical purposes, always present in the air. The spores are wind-borne and are carried to where moisture and temperature conditions are favorable. Under favorable conditions, spores land on surfaces of nondurable wood, and the decay process begins (Figures 6-33 through 6-35).

Insect Attack: Insect damage can occur in standing trees, logs, and in unseasoned and seasoned wood. Insect holes are typically classified as pinholes, grub holes, and powder post holes (created by powder post larvae). When systems of insect tunnels and chambers become

Figure 6-33. Decay voids caused by moisture.
Source: Courtesy of Anthony & Associates.

Figure 6-34. Damage caused by the migration of salts from masonry.
Source: Courtesy of Anthony & Associates.

Figure 6-35. Decay caused by migration from masonry.
Source: Courtesy of Anthony & Associates.

interconnected, they are termed "a network of galleries." Most of the damage done to wood structures is from subterranean termites. Termites are more prevalent in warmer climates, and infestation is greatest in poorly drained areas, and wood close to the ground, because subterranean termites develop and maintain their colonies in the ground. Damage by marine borers to wood foundation piles in salt or brackish waters is practically worldwide. The rapidity of attack depends on local conditions and the kinds of borers present. Untreated woodpiles can be completely destroyed in less than a year in warmer climates. Loss of protective coatings with time can make piles and wood in water susceptible to marine borer attack. Interestingly, pollution of the waterways has provided some protection against marine borers, and as harbors and rivers are cleaned up, marine borers have moved back into these areas with a vengeance (Figure 6-36).

Fire: Characteristic of wood is that it develops a char, which may affect the rate of fire damage on the basis of moisture content and density of wood. Based on the characteristic of a fire, damage may vary across a member and, therefore, requires careful investigation. Wood members subjected to fire develop a layer of char that protects the wood below the char layer by slowing the rate of combustion below the char. In fires where the burning period is relatively brief, many of the wood members will exhibit various stages of char development. Damage to the wood may extend beyond the obvious area of charring (Figures 6-37 and 6-38).

Mechanical Damage: Mechanical damage may result in a loss of load-carrying capacity. Abrasion, saw cuts, and equipment impacts are common mechanisms of mechanical damage. Wood members, such as posts and columns near traffic areas in garages and loading docks, are highly susceptible to mechanical damage resulting from impact. Damage to the wood may range from simple abrasion resulting in reduced cross section to local fractures, split-off corners, or catastrophic failure of individual members. Similar damage can occur during shipment and

Figure 6-36. Insect damage as evidenced by exit holes.
Source: Courtesy of Anthony & Associates.

Figure 6-37. Charred floor system probably requiring full replacement. Source: Courtesy of Anthony & Associates.

Figure 6-38. Measuring the extent of fire damage using a char depth gauge. Source: Courtesy of Anthony & Associates.

improper handling of the members and components during construction and installation, with the damage not being repaired.

Chemical Attack: Chemical attack is less common in buildings than biological deterioration caused by decay or insects. The breakdown of wood substance is typically caused by interaction between the wood constituents reacting to metals or chemicals either in contact with the wood or added to the wood (such as a wood preservative or fire-retardant treatment). Some wood species have chemical extractives that react with metals to discolor or breakdown the wood fibers or the metal. These species are seldom used as construction materials. Ipe is one example of a species with reactive extractives, which often results in staining of the wood when in contact with mild steel. More common are wood preservatives that can corrode steel fasteners. The use of alkaline copper quaternary (ACQ) and copper azole (CuAz) as wood preservatives has been shown to produce corrosion in nongalvanized or stainless steel fasteners and metal plate connectors. Similarly, families of mono-ammonium-phosphate-based fire-retardant treatments used primarily in the 1980s were found to degrade the wood fibers under conditions of elevated temperature and relative humidity. The degradation could result in failure of wood-based panels and solid-sawn lumber.

6.4.7 Factors That Influence Wood Behavior

Moisture: Moisture has a dramatic effect on the properties of wood. The moisture content of wood is the mass of water in the wood expressed as a percentage of the oven-dry mass of the wood. If wood changes moisture content after installation, the resulting change in dimension can cause distortion and twisting. The change in dimension occurs only at moisture contents below the fiber saturation point (approximately 30% moisture content). A decrease below this moisture content will result in an increase in the strength properties of wood, but it will also cause it to shrink, possibly resulting in checks and splits caused by differential shrinkage. Grading of new wood and establishment of reference design values takes into consideration the size and number of checks and splits, as well as knot size and slope of grain. Accordingly, as knot size and slope of grain increase, the reference design values decrease. A change in moisture content above the fiber saturation point does not affect strength or dimensional changes. A reduction in reference design values is equivalent to overloading the structure. A change in moisture content above the fiber saturation point does not affect strength or shrinkage. However, increases below this point result in lower moduli of elasticities and lower permitted reference design values. Thus, an increase in moisture content from the design and installed value up to the fiber saturation point can result in equivalent overloading of the structure.

Effects of Changing Moisture Content. The major effects of changing moisture content of wood are

- *Warp*. Any variation from a plane surface, including bow, crook, cup, or twist as a result of (but not limited to) a change in moisture content. Warp can result in eccentricity of applied load. Deformations caused by bowing and crooking must be considered when evaluating beam, column, and pile strengths.
- *Checks and splits*. Checks and splits can result from differential shrinkage of wood in the radial and tangential directions of the growth rings. Splits and checks are wood failure, and their effect should be considered in the assessment process.

Deterioration: Biological deterioration is typically caused by fungal or insect attack. Bacteria can degrade wood but are typically not a concern with wood structures because decay fungi and insects tend to impact material properties more rapidly than bacteria. All wood is subject to a variety of deterioration mechanisms, the most prevalent being wood-decay fungi, which can ultimately lead to the inability of structural members to perform. Depending on wood species, large timbers will retain moisture internally, frequently causing interior decay with no visible sign of deterioration on the surface of the wood. Moisture absorption through end grain, checks, or holes provides a highly favorable environment for decay fungi to attack the heartwood at the center of a large timber. The heartwood (the inner growth rings of the tree) typically has more decay resistance than the sapwood (the outer growth rings of the tree). However, even the heartwood of naturally durable species, such as chestnut, will decay when exposed to enough moisture. Deterioration is a particular concern where the wood is in contact with the ground or with other materials, such as porous masonry, which may allow for moisture to be absorbed into the wood.

Fungi associated with wood include mildew and stain and decay fungi. Fungi propagate from spores present in the air. Mildew grows on the surface of wood and paint and does not affect the strength of the wood. Stain fungi (not to be confused with moisture stains) penetrate the surface of the wood but do not reduce its strength. Decay fungi, however, break down wood components over time. All types of decay fungi—brown rot, white rot, and dry rot—affect the ability of wood to perform its intended function. Although identifying the specific fungus during wood condition assessment is not essential, identifying the location and extent of deterioration resulting from decay fungi is important.

In general, if the moisture content of the wood is less than 20%, fungi will not be able to grow. Areas with moisture contents between 20% and 30% can support the growth of fungi, but the moisture may not be sufficient to support long-term active decay. Moisture contents between 30% and 40% are highly favorable for active fungal growth and are often an indication of advanced decay, with symptoms that may include internal voids and surface deterioration. Highly saturated wood (with more than 60% to 80% moisture content) typically has insufficient oxygen for fungus to grow; therefore, the decay may be inactive. Insects typically require that moisture be greater than 10% to be active and deteriorate the wood. In addition to favorable moisture content, both fungi and insects require suitable temperature and oxygen, as well as the wood as a food source.

Structural Grade: Wood functions as a structural material in all capacities as beams, columns, diaphragms, bearing plates, or a combination of these to support and distribute vertical and horizontal loads. For solid wood, the ability of wood to function in these capacities is determined by its designated structural grade and associated reference design values. The stress grade is a function of the presence or absence of characteristics that may affect the strength. These include knots, checks, shakes, pitch pockets; slope of grain; and number of annual growth rings per inch. The basic allowable stress is a function of the species, the stress grade, and the size classification (dimension lumber, post, timbers, etc.). Furthermore, the ability of wood to function structurally is related to the moisture content of the wood and its variations, thermal changes either along or cross grain, the direction of application of load to the grain of the wood, existence of preservatives and fire retardants, temperature environment, and the duration of the applied load (intermittent or sustained). Wood is not only susceptible to sudden damage and deterioration, but with time, it can also exhibit permanent deflection, bowing, twisting, and shrinkage, which all have their effect on the capacity of wood to function as a structural material.

Structural grade is not directly impacted by changing moisture content; however, if lumber or timber is graded in the green condition, measured defects that may meet the requirements of a specific grade may not meet the requirements for that grade if the lumber or timber dries and shrinks, thereby reducing the cross-sectional dimensions that are used to determine acceptable defect size. Most commonly, this applies to timbers that are graded in the green condition, and a grade-limiting knot allows for a higher grade than permitted once the timber dries to EMC and has reduced cross section.

Duration of Load and Creep: Duration of load is not a mechanism of deterioration, but it may need consideration when conducting a condition assessment. Duration of load is the length of time a member supports a specific type of load (e.g., impact, wind, seismic, normal, or long term). For a given product under defined load conditions, the load required to produce failure over a long period is less than the load required to produce failure over a short period. Present-day design procedures recognize this condition and allowable stresses are adjusted for duration of load. For some older wood structures, this adjustment was likely not considered in the design and the structures might have become overloaded with time. However, this overload is in the magnitude of approximately 10% and, therefore, is not a serious problem except when other factors affect the allowable stresses and the cumulative effect results in the member not being able to sustain the anticipated loads. Creep is time-dependent deformation under sustained loads. The magnitude is affected by moisture content, temperature, and other factors (Figures 6-39 and 6-40).

Temperature: The mechanical properties of wood are reduced under high and sustained temperatures.

6.4.8 Considerations with Wood Used in Historic Construction

Historic Structures: Prior to conducting an assessment and any repairs, the historic status of an existing structure should be determined, and the impacts of the repairs on that status evaluated. However, regardless of any formally recognized historic status, many structures

Figure 6-39. Deformation of bridge stringers caused by creep.
Source: Courtesy of Anthony & Associates.

Figure 6-40. Deformation of a bridge stringer from wood crushing. Source: Courtesy of Anthony & Associates.

are recognized as places of regional, cultural, and personal significance. As such, every effort should be made to adhere to the Secretary of the Interior's Standards for the Treatment of Historic Properties, which require that the historic character of a property be retained and preserved and prohibit the replacement of intact or repairable historic materials. This includes the character-defining features of a structure, such as exposed framing or roof trusses.

Tension Parallel to Grain Reference Design Values: Research on the tensile strength of dimension lumber led to a reduction in 1976 of reference design values for tension parallel to grain. Prior to the reduction in design values, particularly post-World War II, it was assumed that tensile strength was equal to the modulus of rupture from bending tests. Because most bending tests resulted in failure on the tension face of the specimen, it was assumed that tensile strength governed the capacity of the piece. This was due, in part, to testing of clear wood and lumber that was free of defects permitted in full-size lumber. Testing of full-size lumber revealed that tension parallel to grain strength was, in fact, less than the bending strength, and various reduction factors were applied to bending strength to provide more realistic design properties. Up until the time when reduction factors had been implemented, many structures built to the design loads in tension were found to be overloaded, and some resulted in failure. This behavior was more prevalent in bowstring trusses and glue-laminated timber than other engineered wood products. For structures built in that era that are subjected to a condition assessment, caution should be observed when bowstring trusses or glue-laminated timber are involved.

Bowstring trusses are constructed with a curved top chord and horizontal bottom chord. Their use was common from the 1920s through the 1960s to provide long clear spans. In addition to applying overly optimistic tension design values, bowstring trusses were constructed and installed on the basis of inaccurate code practices and analysis methods. The combination of these factors led to an overstress of these trusses. Assessment of these trusses must account for the initial design and fabrication of the trusses being based on information that has since been refined. To a lesser extent, glued laminated timbers had also relied on the higher tension parallel to grain design values until the 1970s. Although fewer failures have occurred, an assessment of glue-laminated timber prior to the 1970s should consider whether the allowable design values applied to the timber reflects the correct design value for the tension lamination.

6.4.9 Connections

The connection condition assessment includes a determination as to whether the metal, wood, adhesive, or other connector or fastener types that form the connection between wood and other types of structural members has been compromised. Connection capacity can be determined once the wood and connector/fastener conditions are assessed and the connection geometry is established. The initial assessment is typically visual and must be performed by someone who is well-versed in wood connections. Further assessment of the wood members and connectors/fasteners can be performed through NDT testing by someone skilled in these testing techniques to assess the loss of connection capacity.

Connections for structural wood are usually made of metal and/or wood. There are typically three types of sawn lumber connections: dowel-type, metal connectors typically with integral teeth, and shear type. Dowel-type fasteners transfer loads through dowel bearing and bending and include nails, bolts, and screws. Metal connector plates, typically used in lightweight wood trusses, partially penetrate the wood with the integral teeth to transfer the loads near the wood surface. Shear connectors are typically used for heavier loads and include split rings and shear plates. Traditional joinery can be used in timber construction. Notches, holes, and tongues are used in mortise, tenon, and scarf connections to interlock the wood timber members. Most traditional timber connections are in compression and use wood pegs and wedges.

Common connections include hangers for joists and purlins, framing anchors, metal and wood side plates, bent straps, U-straps, bands, clip angles, bearing plates, steel saddles, hanger rods, arch shoes, steel and wood gusset plates, and traditional mortise and tenon connections. Design criteria for wood connections are given in detail in the *National Design Specification (NDS) for Wood Construction with Commentary and NDS Supplement* (AWS 2018) for sawn lumber and *Standard for Design of Timber Frame Structures and*

Commentary, TFEC 1 (TFEC 2019b) for timber structures. To fully assess the connection, the condition of the materials, material properties of the wood members and connectors, and geometry of the connection are required.

6.4.10 Terminology Related to Connector Characteristics and Properties

The physical condition of the connector and the surrounding wood influences the capacity of the connections and should be noted during the field portion of the condition assessment. The following terms are relevant to connectors when conducting a structural condition assessment.

Angle to Grain Load: The angle formed between the axis line of the connector and the longitudinal axis of the wood fibers.

Condition of the Connector: The physical state as determined by observation and with use of simple physical techniques, such as cleaning, scraping, and sounding, can be used to establish the existence of many of the physical conditions included herein.

Condition of Glue: The condition of the glue-laminated or builtup members measured by the percentage of the open joint surface or the total surface of joints that should be glued.

Corrosion: Oxidation of metal caused by a chemical reaction between the metal and the oxygen in the environment; the most common cause of deterioration of unprotected iron and steel products (rust). Both the metal fastener and the wood in the vicinity of the fastener hole can be degraded by metal corrosion.

Cross-Sectional Properties: Actual dimensions and other geometric properties of connectors.

Deformation: Elongation, bending, or twisting of bolts, rods, and other connectors over the elastic limit.

Detailing: Actual diameter and length of bolts, rods, timber connectors, including edge distance, spacing, pitch and end distances of bolts, nails, screws, and timber connectors. Minimum spacing and end-distance requirements are critical for developing the connector design capacity.

Eccentricity: Difference between the center of gravity of the group of connectors used in a connection and the line of action of the load on the connection. If more than one member meets at a connection, then eccentricity of the group of connectors with each member is possible. Eccentricity must be evaluated in three dimensions. The results of eccentricity include tension perpendicular to grain stresses, which can cause members to split.

Elongated Bolt Holes: Condition of oversized bolt holes that develops from an insufficient bearing capacity of the wood owing to applied loads or from incorrectly drilled holes. This can cause unequal load distribution among bolts in a row.

Penetration of Nails and Screws: Actual depth of physical penetration of the connector into the wood.

Splits at Connection: Existence of splits along the fibers at the connections is usually caused by either external loads or wood shrinkage.

Tightness: Physical condition of bolts, nails, rods, screws, and timber connectors that indicate the connection fits snugly. The condition of the entire connection is determined by the tightness of the connectors.

6.4.11 Physical and Mechanical Properties of Connectors

Connectors are typically either metal, wood, or adhesives. The following is a list of connector properties.

Adhesive Shear Strength: The shear strength and resistance to deformation under load for adhesives are based on manufacturer testing results in accordance with ASTM testing procedures. The shear strength is the stress calculated from the maximum load sustained by the adhered wood specimen divided by the bonded area.

Chemical Composition: Indicators of the properties of bolts (including washers and nuts), nails, rods, screws, and timber connectors that can be used to establish corrosion resistance and mechanical properties.

Elongation: Requirement for certain types of timber connectors is the increase in gauge length in a tension test specimen measured prior to fracture as a percentage of the original length. It is a measure of ductility.

Hardness: For threaded fasteners, sufficient hardness ensures that the surfaces of the fasteners have the capacity to transmit load by bearing; it is a specified requirement for nuts and washers and sometimes for bolts and rods.

Lateral Resistance: Resistance to the lateral movement of nails, screws, or bolts that are embedded in wood. Connections in wood develop lateral resistance by fastener bearing. The resistance is dependent on the species (density), fastener diameter and yield strength, the location of the fastener with respect to the edges of connected members, the depth of fastener penetration in the wood, and the orientation of the fastener with respect to the grain of the wood.

Proof Load: Specified load without a measured permanent set; applicable to bolts, rods, and nuts.

Resistance to Delamination: Measure of the integrity of glue joints for structural laminated members expressed as a percentage of total length of the open joint after the joint is subjected to an autoclave test in water at a specified temperature.

Tensile Strength: Stress calculated from the maximum load sustained by a metal connector during a tension test divided by the original cross-sectional area of the specimen; a required physical property for bolts, rods, and some types of timber connectors. For certain materials, an equivalent tensile strength is determined.

Tensile Strength of Fastener: Maximum stress calculated from the maximum load sustained by a metal fastener during a tension test divided by the original cross-sectional area of the fastener. Tensile strength may limit the fastener's ability to develop the strength of the wood when used as a fastener.

Timber Joint Compressive Strength: Maximum compressive load the timber joint can sustain during a full-scale load test.

Timber Joint Tensile Capacity: Maximum tensile load the timber joint can sustain during a full-scale load test.

Hanger Torsional Moment Capacity: Maximum torsional moment the hanger attached to a joist or purlin can sustain during a full-scale load test.

Hanger Vertical Load Capacity: Maximum vertical load a hanger attached to a joist or purlin can sustain during a full-scale load test.

Withdrawal Strength: Resistance to withdrawal of nails and screws in wood. The resistance is dependent on the length and diameter of the fastener, species, and other properties of wood and the location of the nail or screw with respect to the grain of the wood.

Yield Strength: Stress at which a significant increase in strain occurs without a corresponding increase in stress; a property that is important in the evaluation of certain types of timber connectors. For certain materials, an equivalent yield strength is determined.

6.4.12 Mechanisms of Connector Deterioration

Connectors are typically metal, wood, or adhesive and susceptible to attack by moisture, chemicals, mechanical damage, and fire.

Moisture: Primary mechanism for deterioration of metal connectors is moisture, which can lead to corrosion. Metal fasteners can lose their cross section, reducing the capacity of the connector. Moisture can affect the properties of wood, reducing the allowable stresses within the wood and resulting in a reduction in the connection capacity. Some adhesives are sensitive to moisture, reducing their properties and eventually reducing the connection capacity.

Chemical: Metal connectors can be attacked by some chemicals either within the wood members (certain wood preservatives or fire-retardant treatments) or airborne, resulting in corrosion of the connector and reduction of the connector capacity. Adhesives can be adversely affected by chemicals as well, reducing the properties of the adhesive and eventually the connection capacity. Some wood preservatives are known to cause corrosion of nongalvanized or stainless steel fasteners, hangers, and metal plates. The wood does not need to be wet for the fasteners to corrode, although elevated moisture contents accelerate the corrosion. Although wood preservatives have been found to corrode metal fasteners since the 1920s, more recently, water-borne preservatives, including alkaline copper quaternary (ACQ) and copper azole (CuAz), are known to corrode fasteners that have not been hot-dipped-galvanized or made of stainless steel.

Mechanical Damage: Mechanical damage may result in a loss of load-carrying capacity of the connector. Mechanical damage may result in connector section loss through abrasion, cutting, or impact.

Fire: Fire temperature can affect connections through a reduction of material properties of the wood surrounding the connector and the connectors themselves. Wood will char when subjected to fire, but damage to the wood may extend beyond the charred wood areas. Adhesives are temperature sensitive and can change their properties during and after a fire. Connectors, such as metal-toothed plates used in trusses, may become loose during high temperatures because of differential expansion of the metal relative to the wood to which it is attached.

6.4.13 Test Methods

6.4.13.1 Test Methods (Preliminary Assessment Chart). The relative value to the assessment professional of visual, nondestructive, and destructive field testing is depicted in Table 6-30 The chart is a quick reference for the applicability of the certain test methods to determine wood and connection conditions and properties. Subsequent tables list the capabilities and limitations of various test methods in greater detail, along with a list of references that have additional information. Condition assessment of structures is dynamic, and the investigator should recognize that some technologies listed may become unavailable as devices are removed from the market or replaced by newer technology.

Visual, nondestructive, and destructive tests are used to establish the properties and physical conditions of both new wood products and wood used in structures. Most tests are not suitable for testing new products as well as wood products in situ. New test procedures, methodologies, and equipment are developed on an ongoing basis, but most are not commercially available or have gained wide acceptance and should, therefore, be used with a sense of caution. It may be necessary to use a combination of methods for satisfactorily predicting the properties of wood. References cited in

Section 6.4, unless otherwise noted, are included in Appendixes B1 and B3. Additional references are to be found in Appendixes B2 and B4.

Sample Size: Several ASTM standards exist to guide an investigator in determining sample size if a statistically significant sample is desired. Often a statistically significant sample is not available or economically viable. In this case, it is desired to select a sample without bias. This sample could also be a random sample, but not necessarily, as purely random samples in a failure investigation are often not practical given the constraints of the existing structure. As sample size and distribution is determined, care should be exercised to obtain samples from damaged and undamaged and questionable and unquestionable portions of the structure if applicable.

ASTM E122, *Standard Practice for Calculating Sample Size to Estimate, with a Specified Tolerable Error, the Average for a Characteristic of a Lot or Process* (ASTM 2017), can be followed to determine the number of samples required for a desired level of accuracy. The number of samples will depend on

1. Error, E, that can be tolerated in the estimate of the measured value;
2. Preferred probability, or confidence level, that the measured value is within the limit of error, and
3. Estimate of the variation, σ, of the measured value from one unit to another within the population.

Assuming that the measured result is normally distributed and that the population is large, the following equation, based on an assumed 95% confidence interval, can be used to calculate the number of samples, n, for testing

$$n = \left(\frac{1.96\sigma}{E}\right)^2 \tag{6-4}$$

where

n = Number of samples,
σ = Standard deviation of the measured results, and
E = Allowable error in the measured results.

The factor 1.96, based on a desired 95% confidence level, gives a 0.050 probability that the actual value is not within plus or minus the allowable error, E, of the measured property.

Deterministic sampling describes a situation where a set number of samples are available or sampling is constrained by any of the factors discussed herein and only a set number of samples can be taken (access, effects on the element under consideration, budget, etc.). This sampling protocol requires engineering judgment to ensure that a representative estimate of the characteristic under consideration is obtained.

Table 6-30. Relative Assessment Value for Wood and Connection Conditions and Properties.

Factor	Method															
	Species identification	Visual inspection	Remote visual inspection	Visual grading	Digital radioscopy	Infrared thermography	Moisture measurements	Resistance drilling	Stress waves	Surface hardness	Needle penetration	Screw withdrawal	Core drilling	Tension microspecimens	Glue line test	Ground-penetrating radar
Determine wood species	Yes															
Measure moisture content						Limited	Yes									
Locate deterioration		Limited	Limited		Yes	Limited		Yes	Limited	Limited	Limited	Limited				
Quantify deterioration			Limited		Limited			Yes	Limited			Limited	Estimate	Estimate	Limited	
Assess strength				Estimate					Limited	Limited	Limited		Estimate	Estimate		
Determine modulus of elasticity				Estimate					Estimate							
Identify hidden construction defect			Yes		Yes	Limited		Limited								Limited
Locate and assess fasteners		Limited			Yes											Limited

6.4.13.2 Test Methods (Comprehensive Assessment Charts). Testing of specimens from members of a structure to determine chemical and physical properties can provide useful quantitative data. Although removal of specimens may be destructive, careful selection and handling can ensure that the usefulness of the member is not impaired and the damaged areas can be satisfactorily repaired. Tests on specimens provide specific, quantitative information without relying on interpretation or analytical results. Consideration should be given to physical testing of specimens over nondestructive testing methods to determine similar information. It is also advantageous to use the results of physical tests in specimens from members as base data to corroborate nondestructive testing results. Most nondestructive methods require highly qualified and skilled personnel, specialized equipment, and considerable analysis and interpretation of data to provide qualitative and quantitative information.

Available test methods are cited in Tables 6-31 through 6-36. Some test methods listed in the tables have several ASTM standards [e.g., strength tests includes destructive tests methods for small clear specimens (ASTM D143), full-sized lumber (ASTM D198), wood-based panel tests (multiple standards depending on the product and property being tested, and so on], whereas some tests have no ASTM test method but are recognized as appropriate tests for assessing wood (e.g., decay resistance is typically based on visual observations). The tables are not intended to provide the reader with information and details on how to perform and interpret tests but rather to provide information on test methods that are available, what they can accomplish, advantages and limitations of the tests, and a brief statement about qualifications of testing personnel. This information is not exhaustive and is constantly subject to change. Advantages, limitations, and costs can change with time. Accordingly, the references are the most important part of these tables and care has been taken to provide the most current and in-depth information. A perusal of any specific reference will usually lead to an extensive bibliography on the subject. The supplemental references in Appendixes B2 and B4 contain information that can be pursued for additional knowledge.

Table 6-31. Destructive Test Methods for Determining Mechanical and Physical Properties of Wood.

Property	Purpose of test	Principle of operation	Advantages	Limitations
Decay resistance	Evaluation of the natural decay resistance of wood.	Wood samples are exposed in decay chambers to pure cultures of decay fungi. The mass of the sample is determined before and after exposure. The loss of mass is the measure of decay.	Useful in determining the relative values of decay resistance.	The method is qualitative rather than quantitative. Samples from the structure are required.
Elastic properties	Evaluation of Modulus of Elasticity, Module of Rigidity, and Poisson's Ratio.	Standard laboratory tests on clear and straight-grained small pieces of wood.	Useful in determining orthotropic properties for grading and determining allowable properties.	Samples required from the structure. The validity of the results is limited by the number of specimens tested.
Fasteners' withdrawal resistance	Evaluation of resistance to the withdrawal of nails, staples, and screws.	Prisms of wood with fasteners driven into the specimen are tested for fastener withdrawal in a testing machine.	Measures the ability of the fastener to hold adjoining members together.	Samples required from the structure. The validity of the results is limited by the number of specimens tested.

(*Continued*)

Table 6-31. (*Continued*)

Property	Purpose of test	Principle of operation	Advantages	Limitations
Fasteners' mechanical properties	Determination of the hardness, proof load, and tension of threaded fasteners.	Standard laboratory tests on fasteners.	Provides useful information needed for evaluation of connection.	Samples of fasteners must be removed from the structure.
Moisture content	Determination of moisture content as a percentage of oven-dry mass.	The mass of the sample is determined, oven-dried, and the mass is again determined.	Moisture content is one of the most significant variables of wood affecting many properties significantly.	Samples must be taken from the structure. Moisture content will vary with atmospheric conditions and sample locations.
Preservatives	Determination of the quality and effectiveness of wood preservatives.	The mass of the sample is determined, preservative is extracted, and the mass is again determined. Tests associated with moisture content tests in which preservative content and moisture content are separated.	Preservative content is a measure of penetration and therefore the service life of the structure.	Samples must be taken from the structure.

(*Continued*)

Table 6-31. (*Continued*)

Property	Purpose of test	Principle of operation	Advantages	Limitations
Specific gravity	Provides a measure of amount of wood substance present in the sample, providing information on mass, workability, and strength characteristics.	Value can be determined from moisture content tests or by water immersion or flotation methods.	Used to determine dead load of wood and for determining allowable properties.	Specific gravity is an indefinite quantity except for specific conditions under which it is determined. Samples required from the structure.
Strength tests	Determination of the orthotropic mechanical properties of specimens.	Test specimens are subjected to standard evaluation in the testing machine using standard procedures.	Used to determine the structural grade and allowable properties of wood.	Samples must be taken from the structure. Information limited by the number of samples.
Structural grade and allowable properties	Grade and allowable properties needed to assess the strength and life of the structure.	Samples are classified by visual evaluation and from tests of mechanical properties.	Provides useful information for structural assessment.	For good results, a statistical approach must be used, requiring a large number of samples from the structure.

Table 6-32. NDT Methods for Determining Physical Conditions of Wood.

Property	Purpose of test	Principle of operation	Advantages	Limitations
Coring	Coring can be used to determine the extent of decay and the thickness of char from fire damage. Cored samples can be sent to the laboratory to determine density, moisture content, type of decay, and amount of preservative treatment.	An increment borer or cutting plug is used to extract a core. Core holes need to be treated and plugged to prevent decay in the future.	A positive identification of the condition is made. Initial results are rapid.	Coring is time-consuming. In confined areas, it may be difficult or impossible to core. The extent of decay will require additional coring or drilling. Laboratory results are not immediately known.
Drilling	The reduced resistance to drilling indicates a possible pocket of decay. An examination of the drill chips and fines will confirm whether decay has been encountered.	A member is drilled with an electric drill or a brace and bit. The depth when any change in resistance is encountered is recorded. The drill chips and fines are collected. Drill holes should be treated with a preservative.	Initial results are rapid. The procedure is easier than coring. Tool costs are inexpensive.	The extent of the rot pocket may require numerous drill holes. Laboratory results are not immediately known.

(*Continued*)

Table 6-32. (*Continued*)

Property	Purpose of test	Principle of operation	Advantages	Limitations
Moisture meter	Determination of the moisture content of wood. The susceptibility of wood to decay is based in part on its moisture content. The strength of wood is also influenced by moisture. Moisture and its effects determine the fiber saturation point and degree of distortion, twisting, warping, checking, and splitting.	There are two types of moisture meters, dielectric and resistance. For resistance-type meters, probes are driven into the wood and moisture content readings are recorded at varying depths. Dielectric meters do not require the driving of probes into the wood. The meter is simply held against the wood surface, and an average moisture content at a predetermined depth is measured.	Moisture content readings will indicate areas of decay or areas susceptible to decay. High readings indicate paths of moisture from leaking pipes, poor covering, or poor flashing. The meter is small and portable.	Wood treated with creosote and pentachlorophenol can be tested with moisture meters with negligible error. Oxide formulations of CCA types B and C will cause slight errors but will still provide useful results. Meters should not be used with other types of chemical treatments. No meter is accurate above the fiber saturation point (approximately 26%–28% of moisture content). Species and temperature influence the readings. Correction charts are supplied with moisture meters. A moisture meter relative in cost to a good camera must be purchased. Spare probe points and batteries should be stored with the meter.

(*Continued*)

Table 6-32. (*Continued*)

Property	Purpose of test	Principle of operation	Advantages	Limitations
Probing	Probing with an awl or pick will locate surface areas that are soft and decayed. Sound wood that is pried will break into long splinters; decayed wood is brittle and will break into short splinters.	Jab the member with the probe and pry down. Note the resistance to probing and the condition of the pried splinters.	Results are rapid. The test can be done as part of a visual inspection. Costs are low.	Only the surface can be tested. Thick members will have to be tested by other methods to determine the condition of the interior.
Radiograph	Radiographs (X-rays) can be used to determine the condition of highly inaccessible wood members; grain direction, irregularities, decay, splits, and insect damage.	X-rays are taken.	The condition through the piece can be obtained with a nondestructive test. This is desirable in historic structures.	Results are not known until after the film is developed. The tests are costly. Access to both sides of the member is required.

(*Continued*)

Table 6-32. (*Continued*)

Property	Purpose of test	Principle of operation	Advantages	Limitations
Sounding	A poor hammer rebound or a dull or hollow sound indicates possible decay. Loose connections give off a hollow sound and may vibrate when struck with a hammer.	The surface is struck rapidly with changes in sound and rebound noted.	The test is rapid. Results are immediately known. The test can be performed during a visual inspection.	Coring or drilling will be required to determine the depth of decay. Internal decay in a thick member cannot be located.

(*Continued*)

Table 6-32. (*Continued*)

Property	Purpose of test	Principle of operation	Advantages	Limitations
Stress wave propagation	Sonic and ultrasonic techniques exist to detect and spatially quantify areas of advanced decay, insect, and marine borer damage; to determine elastic properties; and to estimate strength properties. Sonic and ultrasonic velocity and spectral content (frequency spectrum representation of stress wave time series) are influenced by density, moisture content, elastic properties, growth characteristics, processing variables, and defects (including deterioration).	Dynamic elastic properties are directly calculated from velocity and wood density. Estimates of static elastic and strength properties can be obtained from measured dynamic elastic properties. Spectral content can be used to estimate strength properties. Estimates of static elastic and strength properties using either velocity or spectral content depend on empirical relationships. Sudden reductions in velocities through the thickness of a member or shifts in spectral content are indicators of void, advanced decay, insect damage, and so on.	Elastic properties can be determined with a high level of accuracy. Excellent methods for initial screening of members for the presence of advanced decay and voids. Some techniques offer immediate on-site estimates of mechanical properties. Most equipment is portable and test times range from quick to moderately long depending on the methods used and desired accuracy. The use of stress wave techniques, in combination with visual and/or sound-and-bore techniques, offers an excellent compromise between test complexity and accuracy of results.	Accurate assessment of elastic properties depends on knowing wood density. Velocity and spectral content are influenced by wood moisture content, temperature, chemical treatment, and orientation of grain and growth rings to the direction of stress wave propagations, which can lead to errors in interpreting test data. Only a modest level of accuracy can be obtained in spatial quantification of decay and voids and in estimation of strength properties.

(*Continued*)

Table 6-32. (*Continued*)

Property	Purpose of test	Principle of operation	Advantages	Limitations
Visual and visual stress grading	A visual inspection is used to determine the appearance and visible condition of the member. If a grade stamp is not visible, then the physical dimensions of the member, size, and location of growth characteristics are used to determine the grade. The cause of failure may be evident by the appearance of the failed members.	Visually examine all exposed surfaces and record all dimensions and conditions found—for example, knots, checks, splits, slope of grain, stains, and connection details. Record all details of fire damaged and failed members. Consider digital audio and video documentation, and/or photographing during a visual inspection.	Results are usually rapid. A general inspection (walk-through) can be done quickly and a trained eye will know where to look for visible damage. A survey will take a longer time. Cost is principally for manpower and not for equipment.	Only visible surfaces can be inspected. Finishes may have to be removed to view hidden surfaces. Load tests may be required for systems not easily exposed. Trained personnel are required.

Table 6-33. Evaluation of Properties of Wood.

	Table 6-31	Table 6-31	Table 6-31	Table 6-31	Table 6-37	Table 6-31	Table 6-31	Table 6-32	Table 6-32	Table 6-31	Table 6-32	Table 6-31	Table 6-32	Table 6-31	Table 6-31
Evaluation procedure / Mechanical and physical property	Bending tests	Compression tests	Density tests	Fatigue tests	Load tests	Moisture tests	Modulus tests	Penetration test	Radiographic	Shear tests	Stress wave propagation	Tension tests	Visual examination	Visual grading	Species identification
Bending strength	•				•										
Classification													•	•	
Compression parallel to grain		•						•			•				
Compression perpendicular to grain		•													
Density		•					•		•						
Fatigue strength				•											
Fiber saturation point						•									

(Continued)

Table 6-33. (*Continued*)

Evaluation procedure / Mechanical and physical property	Bending tests	Compression tests	Density tests	Fatigue tests	Load tests	Moisture tests	Modulus tests	Penetration test	Radiographic	Shear tests	Stress wave propagation	Tension tests	Visual examination	Visual grading	Species identification
	Table 6-31	Table 6-31	Table 6-31	Table 6-31	Table 6-37	Table 6-31	Table 6-31	Table 6-32	Table 6-32	Table 6-31	Table 6-32	Table 6-31	Table 6-32	Table 6-31	Table 6-31
Grade													•	•	
Growth characteristics															
Heartwood													•		
Knots									•				•	•	
Sapwood													•		
Slope of grain		•							•				•	•	
Modulus of elasticity	•						•				•				

(*Continued*)

Table 6-33. *(Continued)*

	Table 6-31	Table 6-31	Table 6-31	Table 6-31	Table 6-37	Table 6-31	Table 6-31	Table 6-32	Table 6-32	Table 6-31	Table 6-32	Table 6-31	Table 6-32	Table 6-31	Table 6-31
Evaluation procedure / Mechanical and physical property	Bending tests	Compression tests	Density tests	Fatigue tests	Load tests	Moisture tests	Modulus tests	Penetration test	Radiographic	Shear tests	Stress wave propagation	Tension tests	Visual examination	Visual grading	Species identification
Shear strength										•					
Species													•		•
Tensile strength parallel to grain												•			
Tensile strength perpendicular to grain												•			

Table 6-34. Evaluation of Physical Conditions of Wood.

	Table 6-32	Table 6-32	Table 6-32	Table 6-32	Table 6-32	Table 6-32	Table 6-31	Table 6-32	Table 6-32	Table 6-32	Table 6-31	Table 6-32	Table 6-32	Table 6-32	Table 6-31	Table 6-31
Evaluation procedure / Physical condition	Coring	Decay tests	Drilling	Glue tests	Moisture meter	Penetration test	Physical measurement	Probing	Radiography	Sounding	Species identification	Stress wave propagation	Termite tests	Visual examination	Wood treatment tests	Visual grading
Adhesive																
Chemical exposure											•			•		
Chemical treatments															•	
Condition of glue				•												
Cross-sectional properties							•									

(Continued)

Table 6-34. (*Continued*)

	Table 6-32	Table 6-32	Table 6-32	Table 6-32	Table 6-32	Table 6-32	Table 6-31	Table 6-32	Table 6-32	Table 6-32	Table 6-31	Table 6-32	Table 6-32	Table 6-32	Table 6-31	Table 6-31
Evaluation procedure / Physical condition	Coring	Decay tests	Drilling	Glue tests	Moisture meter	Penetration test	Physical measurement	Probing	Radiography	Sounding	Species identification	Stress wave propagation	Termite tests	Visual examination	Wood treatment tests	Visual grading
Decay and decay exposure	•	•	•		•	•		•	•		•	•		•		
Deflection							•							•		
Fire retardants and preservatives															•	
Fungal damage								•				•		•		
Insect infestation and animal attack								•	•			•	•	•		

(*Continued*)

Table 6-34. (*Continued*)

	Table 6-32	Table 6-32	Table 6-32	Table 6-32	Table 6-32	Table 6-32	Table 6-31	Table 6-32	Table 6-32	Table 6-32	Table 6-31	Table 6-32	Table 6-32	Table 6-32	Table 6-31	Table 6-31
Evaluation procedure / Physical condition	Coring	Decay tests	Drilling	Glue tests	Moisture meter	Penetration test	Physical measurement	Probing	Radiography	Sounding	Species identification	Stress wave propagation	Termite tests	Visual examination	Wood treatment tests	Visual grading
Moisture and its effect																
Pattern of exposure					•									•		
Effects of changing moisture content																
Checks and end splits									•					•		•
Shrinkage														•		

(*Continued*)

Table 6-34. (*Continued*)

	Table 6-32	Table 6-32	Table 6-32	Table 6-32	Table 6-32	Table 6-32	Table 6-31	Table 6-32	Table 6-32	Table 6-32	Table 6-31	Table 6-32	Table 6-32	Table 6-32	Table 6-31	Table 6-31
Evaluation procedure / Physical condition	Coring	Decay tests	Drilling	Glue tests	Moisture meter	Penetration test	Physical measurement	Probing	Radiography	Sounding	Species identification	Stress wave propagation	Termite tests	Visual examination	Wood treatment tests	Visual grading
Twisting							•							•		
Warping (bowing, crooking, and cupping)							•							•		•
Moisture content					•				•							
Physical damage	•		•				•	•	•					•		
Rot	•		•			•		•	•			•		•		

(*Continued*)

Table 6-34. (*Continued*)

	Table 6-32	Table 6-32	Table 6-32	Table 6-32	Table 6-32	Table 6-32	Table 6-31	Table 6-32	Table 6-32	Table 6-32	Table 6-31	Table 6-32	Table 6-32	Table 6-32	Table 6-31	Table 6-31
Evaluation procedure / Physical condition	Coring	Decay tests	Drilling	Glue tests	Moisture meter	Penetration test	Physical measurement	Probing	Radiography	Sounding	Species identification	Stress wave propagation	Termite tests	Visual examination	Wood treatment tests	Visual grading
Seasoning					•									•		
Temperature environment														•		
Time load duration (time creep)							•							•		
Weathering														•		

Table 6-35. Evaluation of Properties of Connectors of Wood.

	Table 6-12	Table 6-32	Table 6-31	Table 6-31	Table 6-12	Table 6-31	Table 6-12	Table 6-31	Table 6-31
Evaluation procedure / Chemical and physical properties	Chemical analysis	Delamination tests	Fastener mechanical tests	Hanger tests	Hardness tests	Lateral resistance tests	Tension tests	Timber joint tests	Withdrawal tests
Bolts and rods (including lag bolts, washers, and nuts)									
Chemical composition	•								
Hardness			•		•				
Proof load			•						
Tensile strength			•						
Joist and purlin hangers									
Torsional moment capacity				•					
Vertical load capacity				•					
Resistance to delamination		•							
Timber connectors (split rings, toothed rings, shear plates)									
Chemical composition	•								
Elongation							•		

(*Continued*)

Table 6-35. (*Continued*)

	Table 6-12	Table 6-32	Table 6-31	Table 6-31	Table 6-12	Table 6-31	Table 6-12	Table 6-31	Table 6-31
Evaluation procedure / Chemical and physical properties	Chemical analysis	Delamination tests	Fastener mechanical tests	Hanger tests	Hardness tests	Lateral resistance tests	Tension tests	Timber joint tests	Withdrawal tests
Tensile strength							•		
Timber joint compressive strength								•	
Timber joint tensile strength								•	
Yield strength							•		
Wood nails									
Chemical composition	•								
Lateral resistance						•			
Withdrawal strength								•	•
Wood screws (including lag screws and spiral dowels)									
Chemical composition	•								
Lateral resistance						•			
Withdrawal strength								•	•

Table 6-36. Evaluation of Physical Conditions of Connectors of Wood.

	Tables 6-14 and 6-32	Tables 6-14 and 6-32	Tables 6-14 and 6-32
Physical condition \ Evaluation procedure	Physical measurements	Visual examination	Corrosion rate
Angle of axis of connector to grain	•		
Condition of connector		•	
Condition of glue		•	
Corrosion		•	•
Cross-section properties	•		
Deformation	•	•	
Dimensions	•		
Eccentricity	•	•	
Elongated bolt holes	•	•	
Nonbinding condition		•	
Penetration of nails and screws	•		
Splits at connection		•	
Tightness		•	

6.5 LOAD TEST METHODS

The condition of the building may require member component or system load testing to understand the strength of the structure relative to the demand on the structure. Sections 6.1 through 6.4 of this document have discussed the assessment of typical building materials. Based on the information collected during the material assessments, field testing may be required to determine the building component's or system's strength capacity.

Available load test methods and data reduction are cited in Table 6-37. The table is not intended to provide the reader with information and details on how to perform and interpret tests but rather to provide information on test methods that are available, what they can accomplish, advantages and limitations of the tests and a brief statement about qualifications of testing personnel. This information is not exhaustive and constantly subject to change. Advantages, limitations, and costs can change with time. The supplemental references in Appendixes B2 and B4 contain formation that can be perused for additional knowledge.

Table 6-37. Test Methods for Load Testing on Buildings

Test method	Purpose of test	Principle of operation	Capabilities	Limitations
Analysis and interpretation of test data	Limits states criteria, proof loads and structure reliability	Mathematical analysis of load test data using statistical interpretation when applicable	Useful to determine the load capacities of members and systems. A qualified engineer is required to perform and interpret the analysis and data.	Limited by the amount and type of test data recorded.
Dynamic load testing	Mainly applicable to building systems	An existing force is applied to a member or structural system and the structure instrumented to measure dynamic response.	Useful to identifying defects but also provides data for future testing comparison. A qualified engineer is required to design and supervise the load tests.	Limited by costs and requires specially made equipment to apply excited forces.
Monitoring dynamic response	Usually, permanent systems to monitor seismic activity	Instrument critical location and members. Use velocity transducers, accelerometers, or seismometers.	The structure signature is obtained for seismic loads. A qualified engineer is required.	This test is not very applicable in identifying specific deterioration problems.

(*Continued*)

Table 6-37. *(Continued)*

Test method	Purpose of test	Principle of operation	Capabilities	Limitations
Monitoring strain/stress	Strains are recorded under service loads and sometimes for static load tests.	Instrument members and measure strain at critical locations with the use of strain gauges.	Provides useful information to pin-point overstressed members. Typically monitored over a long period of time. A qualified engineer is required.	Limited by cost and the number of stations required to obtain meaningful data.
Static load tests	In-situ load to predict how members or systems will perform. Can generally determine only load capacity based on bending.	Specific and identifiable loads corresponding to multiples of the service load are applied and deflections/strains are recorded for the severe load applications.	Can provide actual proof loads for the component or system. It is the most highly used test method. A qualified engineer is required.	Limited by cost.

APPENDIX A

STRUCTURAL CONDITION ASSESSMENT REPORT (NOT PART OF GUIDELINE)

1 EXECUTIVE SUMMARY (OPTIONAL)

2 INTRODUCTION

2.1 PURPOSE

2.1.1 Determine current conditions/deficiencies
- (a) Life safety
- (b) Performance
- (c) Serviceability

2.1.2 Planning
- (a) Maintenance
- (b) Repairs
- (c) Budgeting

2.1.3 Change of owner

2.1.4 Change of occupancy

2.1.5 Alterations or additions

2.1.6 Code compliance
- (a) Life safety
- (b) Fire safety
- (c) Strengthening for lateral forces

2.1.7 Adaptive reuse, rehabilitation, restoration

2.1.8 Historic preservation

2.1.9 Distress, failure, or damage
- (a) System or material failure
- (b) Cracking, bulging
- (c) Water intrusion: rot, freeze–thaw damage, ice, corrosion

(d) Fire
(e) Flood
(f) Wind: storm, tornado, hurricane
(g) Blast, impact, vibration
(h) Seismic event
(i) Subsidence, differential settlement, sinkhole, heaving, swelling soils
(j) Connection failure
(k) Deterioration, erosion, weathering
(l) Insect, rodent, or bird infestation
(m) Others

2.2 SCOPE OF INVESTIGATION AND ASSESSMENT

2.2.1 Cursory assessment/screening
2.2.2 Preliminary assessment
2.2.3 Detailed assessment
2.2.4 Survey
(a) Corroboration of existing drawings
(b) Measured drawings, measurement of structural members
(c) Field evaluation of conditions (e.g., visual, probing)
(d) Identification of problem areas
(e) Record of observations (prints, matrix, photos, tapes)

2.2.5 Testing
(a) Nondestructive
(b) Destructive
(c) Load test

2.3 METHODS AND TECHNIQUES

2.3.1 Visual—including binoculars, magnifying glass, borescope, and fiber optics
2.3.2 Photography, X-ray, infrared thermography, ultrasonic, impulse-echo, impulse-radar
2.3.3 Tapes—oral, video
2.3.4 Drawings and sketches
2.3.5 Measurement
2.3.6 Investigation procedure and tests used

2.4 MEETINGS

3 DESCRIPTION OF THE STRUCTURE

3.1 GENERAL

3.1.1 Type of architecture
3.1.2 Type of structure

3.1.3 Materials
- (a) Masonry—stone, brick, CMU, clay tile, terra cotta, adobe, rammed earth
- (b) Wood—logs, hewn, sawn, laminated, treated, connections
- (c) Metal—iron, steel, aluminum, copper, bronze, lead
- (d) Concrete—placed, precast, plain, reinforced, prestressed

3.2 History

3.2.1 Dates of construction, alteration, and repair
3.2.2 Uses and occupancy—alterations, equipment, vibration, wear
3.2.3 Environmental conditions—weather, heat, cold, chemicals, food products
3.2.4 Site conditions
3.2.5 Unusual loadings

3.3 COLLECTED DATA

3.3.1 Available drawings, specifications, calculations, reports
3.3.2 Insurance descriptions, tax maps, deeds, building permits
3.3.3 Alterations documentation
3.3.4 Photographs and tapes
3.3.5 Measured drawings from the survey
3.3.6 Interviews of people familiar with the building

4 DISCUSSION OF SITE VISITS
4.1 OVERVIEW—may be entire engagement
4.2 SURVEY

4.2.1 Materials
4.2.2 Real or inferred systems
4.2.3 Dimensions
4.2.4 Deflections, distortions, deterioration
4.2.5 Identification of problem areas
4.2.6 Record of data
- (a) Sketches and drawings
- (b) Notes
- (c) Photographs, X-rays, infrared scans, etc.
- (d) Tapes—oral, video

4.3 OBSERVATIONS AND THEIR SIGNIFICANCE

- (a) General
- (b) Limitations and qualifications
- (c) Need for immediate repairs

5 PRELIMINARY OFFICE ANALYSIS

5.1 COMPUTATIONAL ANALYSIS

(a) Determine live, dead, wind, seismic, and other appropriate loads
(b) Structure geometry
(c) Material properties; allowable working or ultimate strength parameters
(d) Methods of analysis
(e) Calculate load resistance and deformation
(f) Determine the need for material samples, member samples

5.2 IDENTIFICATION OF TEST AREAS, APPROPRIATE TESTS

5.3 CODE CONFORMANCE (PAST AND PRESENT)

6 TEST PROGRAM

6.1 NONDESTRUCTIVE METHODS

6.2 DESTRUCTIVE METHODS

6.3 LOAD TEST

6.3.1 System or component

6.3.2 Methods

6.3.3 Instrumentation

6.4 RESULTS

7 FINAL ANALYSIS (BASIS FOR EVALUATION)

8 INPUT FROM OTHER DISCIPLINES

9 SUMMARY OF THE STUDY

10 CONCLUSIONS

11 RECOMMENDATIONS

12 APPENDIXES

(a) Survey information
(b) Record drawings
(c) Photographs
(d) Test data and reports
(e) Computations
(f) References

APPENDIX B1
STRUCTURAL MATERIAL ASSESSMENT—REFERENCES

The following references have been organized by specific material or general category.

CONCRETE

ACI (American Concrete Institute). 1971. *Hardened concrete: Physical and mechanical aspects*. Monograph No. 6. Farmington Hills, MI: ACI.

ACI. 2008. *Guide for conducting a visual inspection of concrete in service*. ACI PRC 201.1-08. Farmington Hills, MI: ACI.

ACI. 2013a. *Code requirements for load testing of existing concrete structures and commentary*. ACI 437.2-13. Farmington Hills, MI: ACI.

ACI. 2013b. *Guide to selecting protective treatments for concrete*. ACI PRC 515.2-13. Farmington Hills, MI: ACI.

ACI. 2013c. *Report on nondestructive test methods for evaluation of concrete in structures*. ACI PRC 228.2-13. Farmington Hills, MI: ACI.

ACI. 2016. *Guide to durable concrete*. ACI PRC 201.2-16. Farmington Hills, MI: ACI.

ACI. 2017. *Report on service life prediction*. ACI 365.1R-17. Farmington Hills, MI: ACI.

ACI. 2018. *Specifications for repair of concrete in buildings*. ACI 563-18. Farmington Hills, MI: ACI.

ACI. 2019a. *Building code requirements for structural concrete and commentary*. ACI 318-19. Farmington Hills, MI: ACI.

ACI. 2019b. *Guide for assessment of concrete structures before rehabilitation*. ACI PRC 364.1. Farmington Hills, MI: ACI.

ACI. 2020. *Guide for assessment and surface preparation for application of protective systems for concrete*. ACI PRC 515.3-20. Farmington Hills, MI: ACI.

ACI. 2021. *Code requirements for assessment, repair, and rehabilitation of existing concrete structures and commentary*. ACI 562-19. Farmington Hills, MI: ACI.

ASTM International. 2006 *Significance of test and properties of concrete & concrete-making materials*. ASTM STP169D. West Conshohocken, PA: ASTM.

ASTM. 2019. *Standard determining density of structural lightweight concrete*. ASTM C567. West Conshohocken, PA: ASTM.

Cady, P. D., and R. E. Weyers. 1983. *Chloride penetration and the deterioration of concrete bridge decks*. West Conshohocken, PA: ASTM.

Carrier, R. F., D. C. Pu, and P. D. Cady. 1975. "Moisture distribution in concrete bridge decks and pavements." In *Durability of concrete*, 169–190. SP-47. Farmington Hills, MI: ACI.

Evans, R. H., and M. S. Marathe. 1968. "Microcracking and stress–strain curve for concrete in tension." *Mater. Struct. Res. Test.* 1: 61–64. Paris: RILEM.

Gerwick, B. C., Jr. 1991. "The global advance—Emerging opportunities of home and abroad." *PCI J.* 36 (6): 32–37.

Gopalaratnam, V. S., and S. P. Shah. 1985. "Softening response of plain concrete in direct tension." *ACI J.* 82 (27): 310–323.

Hughes, B. P., and G. P. Chapman. 1966. "The complete stress–strain curve for concrete in direct tension." *Mater. Struct. Res. Test.* 30: 95–97.

Litvan, G. G. 1980. "Freeze–thaw durability of porous building materials." In *Durabilities of building materials and components*, edited by P. J. Sereda and G. G. Litvan, 455–463. ASTM STP 691. West Conshohocken, PA: ASTM.

Mather, B. 1968. "Cracking induced by environmental effects." In *Causes, mechanism, and control of cracking in concrete*, 67–73. Farmington Hills, MI: ACI.

Mather, B. 1989. "Concrete strength-quality assurance, structural materials." In *Proc., Structures Congress '89*, edited by J. F. Orofino, 442–450. Reston, VA: ASCE.

Plowman, J. M. 1956. "Maturity and the strength of concrete." *Mag. Concr. Res.* 8 (22): 13–22.

Shah, S. P., and G. Winter. 1968. "Inelastic behavior and fracture of concrete." In *Causes, mechanism, and control of cracking in concrete*, 5–28. SP-20. Farmington Hills, MI: ACI.

SHRP (Strategic Highway Research Program). 1993. *Distress identification manual for the long-term pavement project*. SHRP-P-338 (FHWA-FA-94-086). Washington, DC: National Research Council.

Stutzman, P. E., and J. R. Clifton. 1994. *Diagnosis of causes of concrete deterioration in the MLP-7A parking garage*. NISTIR 5492. Gaithersburg, MD: NIST.

Weyers, R. E., and D. G. Smith. 1989. "Chloride diffusion constant for concrete, structural materials." In *Proc., Structures Congress '89*, edited by J. F. Orofino, 106–115. Reston, VA: ASCE.

Winter, G. 1961. "Properties of steel and concrete and the behavior of structures." *Trans. ASCE* 126: 1054–1100.

Wong, H. E., and R. G. Drysdale. 1983. "Compression characteristics of concrete block masonry prisms." In *Masonry: Research, Application, and Problems*, edited by J. C. Groyan and J. T. Conway, 167–177. ASTM STP 871. West Conshohocken, PA: ASTM.

METALS

AA (Aluminum Association). 2020. *Aluminum design manual.* Arlington, VA: AA.

AISC (American Institute of Steel Construction). 1989. *Manual of steel construction.* Chicago: AISC.

AISC. 2022. *Steel design guide 27: Structural stainless steel.* Chicago: AISC.

AISC. 2023. *Steel construction manual.* 16th ed. Chicago: AISC.

AISI (American Iron and Steel Institute). 2017. *Cold-formed steel design manual.* AISI D100. Washington, DC: AISI.

Ferris, H. W. 1953. *Historical record, dimensions and properties, rolled shapes, steel and wrought iron beams and columns, as rolled in U.S.A., period 1873–1952 with source noted.* Chicago: American Institute of Steel Construction.

Haynie, F. H. 1980. "Theoretical air pollution and climate effects on materials confirmed by zinc corrosion data." In *Durabilities of building materials and components*, edited by P. J. Sereda and G. G. Litvan, 157–175. ASTM STP 691. West Conshohocken, PA: ASTM.

MASONRY

ASTM International. 2019. *Standard specification for mortar for unit masonry.* ASTM C270. West Conshohocken, PA: ASTM.

ASTM. 2020. *Standard specification for grout for masonry.* ASTM C476. West Conshohocken, PA: ASTM.

ASTM. 2022. *Standard guide for selection, design, and installation of dimension stone attachment systems.* ASTM C1242-22C. West Conshohocken, PA: ASTM.

Bailey, W. G., J. H. Matthys, and J. E. Edwards. 1990. "Initial rate of absorption of clay brick considering both bed surfaces in the as received condition and after outside exposure." In *Masonry: Components to assemblages*, edited by J. H. Matthys, 5–26. ASTM STP 1063. West Conshohocken, PA: ASTM.

BIA (Brick Industry Association). 2019. "Accommodating expansion of brickwork." *Tech. Note 18a*. Reston, VA: BIA.

Hamid, A. A., and R. G. Drysdale. 1980. "Effect of strain gradient on tensile strength of concrete blocks." In *Masonry: Materials, properties, and performance*, edited by J. G. Borchelt, 57–65. ASTM STP 778. West Conshohocken, PA: ASTM.

Hamid, A. A., and M. Schuller 2019. *Assessment and retrofit of masonry structures*. Longmont, CO: The Masonry Society.

Hatzinikolas, M., J. Longworth, and J. Warwaruk. 1980. "Properties of ready-mixed mortar." In *Masonry: Materials, properties, and performance*, edited by J. G. Borchelt, 15–26. ASTM STP 778. West Conshohocken, PA: ASTM.

Haver, C. A., D. W. Keeling, S. Somayaji, D. Jones, et al. 1990. "Corrosion of reinforcing steel and wall ties in masonry systems." In *Masonry: Components to assemblages*, edited by J. H. Mathys, 173–193. ASTM STP 1063. West Conshohocken, PA: ASTM.

Isberner, A. W., Jr. 1980. "Grout for reinforced masonry." In *Masonry: Materials, properties, and performance*, edited by J. G. Borchelt, 38–56. ASTM STP 778. West Conshohocken, PA: ASTM.

Isberner, A. W. 1983. "A test method for measuring the efflorescence potential of masonry mortars." In *Masonry: Research, application, and problems*, edited by J. C. Grogan and J. T. Conway, 27–37. ASTM STP 871. West Conshohocken, PA: ASTM.

Litvan, G. G. 1980. "Freeze–thaw durability of porous building materials." In *Durabilities of building materials and components*, edited by P. J. Sereda and G. G. Litvan, 455–463. ASTM STP 691. West Conshohocken, PA: ASTM.

Plumber, H. C. 1950. *Brick and tile engineering: Handbook of design*. Washington, DC: Structural Clay Products Institute.

Schuller, M. 2003. "Nondestructive testing and damage assessment of masonry structures." In *Progress in structural engineering and materials*. Hoboken, NJ: Wiley.

Stockbridge, J. G. 1980. "Evaluation of terra cotta in-service structures." In *Durabilities of building materials and components*, edited by P. J. Sereda and G. G. Litvan, 218–230. ASTM STP 691. West Conshohocken, PA: ASTM.

Structural Clay Products Institute. 1969. Principles of tile engineering and principles of brick. engineering and recommended practice for engineered brick masonry. Washington, DC: Structural Clay Products Institute.

Suprenant, B. A., and M. P. Schuller. 1994. *Nondestructive evaluation and testing of masonry structures*. Addison, IL: The Aberdeen Group.

TMS (The Masonry Society). 2022. *Building code requirements for masonry structures*. TMS 402. Longmont, CO: TMS.

TMS. 2022. *Specification for masonry structures*. TMS 602. Longmont, CO: TMS.

WOOD

AITC (American Institute of Timber Construction). 2012. *Timber construction manual*. New York: Wiley.

ANSI/AITC (American National Standards Institute/American Institute of Timber Construction). 2007. *American national standard for wood products–Structural glued laminated timber*. A190.1-22. New York: ANSI.

APA (APA–The Engineered Wood Association). 2019. *Engineered wood construction guide*. Tacoma, WA: APA.

APA. 2020. *Glued laminated beam design tables*. APA S475. Tacoma, WA: APA.

APT (Association for Preservation Technology). 2004. Condition assessment of timber using resistance drilling and digital radioscopy. *APT Bull: J. Preserv. Tech.* Springfield, IL: APT.

AWC (American Wood Council). 2018a. *National design specification for wood construction*. Leesburg, VA: AWC.

AWC. 2018b. *National design specification supplement*. Leesburg, VA: AWC.

AWPA (American Wood Protection Association). 2002. *Plywood, fire retardant treatment by pressure processes*. AWPA C27. Washington, DC: AWPA.

AWPA. 2003. *Structural lumber fire-retardant treatment by pressure processes*. AWPA C20. Washington, DC: AWPA.

Breyer, D. E. 2019. *Design of wood structures ASD/LRFD*. 8th ed. New York: McGraw-Hill.

Gurfinkel, G. 1981. *Wood engineering*. New Orleans: Southern Forest Products Association.

Hoyle, R. J., and F. E. Woeste, Jr. 1989. *Wood technology in the design of wood structures*. 5th ed. Ames, IA: Iowa State University.

NPA (National Particleboard Association). 1982. *Standard for particle board decking for factory-building housing*. Silver Spring, MD: NPA.

Ross, R. 2021. *Wood handbook: Wood as an engineering material*. Washington, DC: USDA Forest Service, US Dept. of Agriculture.

TFEC (Timber Frame Engineering Council). 2019a. *Guide to structural evaluation of existing timber structures and structural timber assessment using nondestructive evaluation*. Alstead, NH: TFEC.

TFEC. 2019b. *Standard for design of timber frame structures and commentary*. TFEC 1. Alstead, NH: TFEC.

TPI (Truss Plate Institute). 2014. *National design standard for metal plate connected wood truss construction*. Waldorf, MD: TPI.

GENERAL

ASCE. 2017. *Seismic evaluation and retrofit of existing buildings*. ASCE 41-17. Reston, VA: ASCE.

ASCE. 2022. *Minimum design loads and associated criteria for buildings and other structures*. ASCE 7-22. Reston, VA: ASCE.

ASTM International. 2004 (2018). *Standard guide for conducting flexural tests on beams and girders for building construction*. ASTM E529. West Conshohocken, PA: ASTM.

ASTM. 2019. *Standard test method for static load testing of framed floor or roof diaphragm constructions for buildings*. ASTM E455. West Conshohocken, PA: ASTM.

ASTM. 2022. *Standard practice for calculating sample size to estimate, with a specified tolerable error, the average for a characteristic of a lot or process*. ASTM E122. West Conshohocken, PA: ASTM.

BOCA (Building Officials and Code Administrators). 1996. *BOCA national building code*. Country Club Hills, IL: BOCA.

Evans, R. M. 1993. *Polyurethane sealants—Technology and applications*. Lancaster, PA: Technomic Publishing.

Garden, G. K. 1980. "Design determines durability." In *Durabilities of building materials and components*, edited by P. J. Sereda and G. G. Litvan, 31–37. ASTM STP 691. West Conshohocken, PA: ASTM.

ICBO (International Conference of Building Officials). 1997. *Uniform building code*. Whittier, CA: ICBO.

ICC (International Code Council). 2018. *International building code*. Washington DC: ICC.

SBCC (Southern Building Code Congress). 1997 *Standard building code*. Birmingham, AL: SBCC.

APPENDIX B2
STRUCTURAL MATERIAL ASSESSMENT SUPPLEMENTARY—REFERENCES

The following references have been organized by specific material or general category.

CONCRETE

Alexander, A. M. 1992. "Accuracy of predicting in-situ compressive strength of deteriorated concrete seawall by NDT methods." In *Proc., Nondestructive Evaluation of Civil Structures and Materials*, edited by B. A. Suprenant, J. L. Noland, and M. P. Schuller, 68–82. Boulder, CO: University of Colorado.

Alexander, A. M., and H. J. Thornton, Jr. 1988. "Developments in ultrasonic pitch-catch and pulse-echo for measurements in concrete." In *Nondestructive testing*, edited by H. S. Lew, 21–40. SP-112. Farmington Hills, MI: American Concrete Institute.

Bruns, S. R., and C. C. Higgins. 1992. "Evaluation and testing of a closed spandrel reinforced concrete arch bridge." In *Proc., Nondestructive Evaluation of Civil Structures and Materials*, edited by B. A. Suprenant, J. L. Noland, and M. P. Schuller, 395–410. Boulder, CO: University of Colorado.

Buchanan, D. J. 1992. "Use of hydrophones to monitor prestressed concrete pipe." In *Proc., Nondestructive Evaluation of Civil Structures and Materials*, edited by B. A. Suprenant, J. L. Noland, and M. P. Schuller, 427–435. Boulder, CO: University of Colorado.

Bungey, J. H. 1984. "The influence of reinforcement on ultrasonic pulse velocity testing." In *In situ/nondestructive testing of concrete*, edited by V. M. Malhotra, 229–246. SP-82. Farmington Hills, MI: American Concrete Institute.

Cantor, T. R. 1984. "Review of penetrating radar as applied to nondestructive evaluation of concrete." In *In situ/nondestructive testing of concrete*, edited by V. M. Malhotra, 581–601. ACI SP-82. Farmington Hills, MI: American Concrete Institute.

Carino, N. J. 1984. "Laboratory study of flaw detection in concrete by the pulse-echo method." In *In situ/nondestructive testing of concrete*, edited by V. M. Malhotra, 557–579. SP-82. Farmington Hills, MI: American Concrete Institute.

Carlsson, M., I. R. Eeg, and P. Jahren. 1984. "Field experience in the use of the "break-off tester"." In *In situ/nondestructive testing of concrete*, edited by V. M. Malhotra, 277–292. SP-82. Farmington Hills, MI: American Concrete Institute.

Cohen, J. S., and A. E. N. Osborn. 1992. "Strain gage monitoring for the queens midtown viaduct." In *Proc., Nondestructive Evaluation of Civil Structures and Materials*, edited by B. A. Suprenant, J. L. Noland, and M. P. Schuller, 453–470. Boulder, CO: University of Colorado.

Dahl-Jorgensen, E., and R. Johansen. 1984. "General and specialized use of the break-off concrete strength testing method." In *In situ/nondestructive testing of concrete*, edited by V. M. Malhotra, 293–308. SP-82. Farmington Hills, MI: American Concrete Institute.

Davis, A. G., and J. Paquet. 1992. "Monitoring bridge performance using NDE techniques." In *Proc., Nondestructive Evaluation of Civil Structures and Materials*, edited by B. A. Suprenant, J. L. Noland, and M. P. Schuller, 227–240. Boulder, CO: University of Colorado.

Dilly, R. L., and W. L. Vogt. 1988. "Pullout test, maturity, and PC spreadsheet software." In *Nondestructive testing*, edited by H. S. Lew, 193–218. SP-112. Farmington Hills, MI: American Concrete Institute.

Erlin, B. 1990. "The magic of investigative petrography: The practical basis for resolving concrete problems." In *Petrography applied to concrete and concrete aggregates*, edited by B. Erlin and D. Stark, 171–181. ASTM STP 1061. West Conshohocken, PA: ASTM.

Figg, J. W., A. Sheehan, and H. N. Tomsett. 1992. "Assessment of composite concrete floors using a combination of sonic and ultrasonic non-destructive test methods." In *Proc., Nondestructive Evaluation of Civil Structures and Materials*, edited by B. A. Suprenant, J. L. Noland, and M. P. Schuller, 145–160. Boulder, CO: University of Colorado.

Ghorbanpoor, A., C. H. McGogney, and Y. P. Virmani. 1992. "Recent developments in NDE of concrete bridge structures." In *Proc., Nondestructive Evaluation of Civil Structures and Materials*, edited by B. A. Suprenant, J. L. Noland, and M. P. Schuller, 247–261. Boulder, CO: University of Colorado.

Harrell, T. R. 1988. "Nondestructive testing use in evaluating the form removal time for a large diameter tunnel concrete lining." In *Nondestructive testing*, edited by H. S. Lew, 153–164. SP-112. Farmington Hills, MI: American Concrete Institute.

Maji, A. K., and S. P. Shah. 1988. "Application of acoustic emission and laser holography to study microfracture in concrete." In *Nondestructive testing*, edited by H. S. Lew, 83–109. SP-112. Farmington Hills, MI: American Concrete Institute.

Malhotra, V. M. 1984. "In situ/nondestructive testing of concrete—A global review." In *In situ/nondestructive testing of concrete*, edited by V. M. Malhotra, 1–16. SP-82. Farmington Hills, MI: American Concrete Institute.

Mather, K. 1966. "Hardened concrete-petrographic examination." In *Significance of tests and properties of concrete and concrete making materials*, 125–143. ASTM STP 169A. West Conshohocken, PA: ASTM.

Mielenz, R. C. 1966. "Concrete aggregate-petrographic examination." In *Significance of tests and properties of concrete and concrete making materials*, 381–403. ASTM STP 169A. West Conshohocken, PA: ASTM.

Mlakar, P. F., R. E. Walker, B. R. Sullivan, and V. P. Chiarito. 1984. "Acoustic emission behavior of concrete." In *In situ/nondestructive testing of concrete*, edited by V. M. Malhotra, 619–637. SP-82. Farmington Hills, MI: American Concrete Institute.

Nast, T. E., R. Zoughi, and P. S. Nowak. 1992. "Preliminary results of microwave reflectometry as a nondestructive tool for studying concrete properties." In *Proc., Nondestructive Evaluation of Civil Structures and Materials*, edited by B. A. Suprenant, J. L. Noland, and M. P. Schuller, 339–344. Boulder, CO: University of Colorado.

Nazarian, S., and M. Baker. 1992. "A new nondestructive testing device for comprehensive pavement evaluation." In *Proc., Nondestructive Evaluation of Civil Structures and Materials*, edited by B. A. Suprenant, J. L. Noland, and M. P. Schuller, 437–452. Boulder, CO: University of Colorado.

Ohtsu, M. 1988. "Diagnostics of cracks in concrete based on acoustic emission." In *Nondestructive testing*, edited by H. S. Lew, 63–82. SP-112. Farmington Hills, MI: American Concrete Institute.

Ohtsu, M. 1992. "Current NDE developments of concrete structures and materials in Japan." In *Proc., Nondestructive Evaluation of Civil Structures and Materials*, edited by B. A. Suprenant, J. L. Noland, and M. P. Schuller, 1–16. Boulder, CO: University of Colorado.

Olson, L. D., D. A. Sack, and G. C. Phelps. 1992. "Sonic NDE of bridges and other concrete structures." In *Proc., Nondestructive Evaluation of Civil Structures and Materials*, edited by B. A. Suprenant, J. L. Noland, and M. P. Schuller, 279–295. Boulder, CO: University of Colorado.

Pla, G., and M. O. Eberhard. 1992. "Applications of imaging technology to the nondestructive evaluation of reinforced concrete." In *Proc., Nondestructive Evaluation of Civil Structures and Materials*, edited by B. A. Suprenant, J. L. Noland, and M. P. Schuller, 99–112. Boulder, CO: University of Colorado.

Sansalone, M., and N. J. Carino. 1988. "Laboratory and field studies of the impact-echo method for flaw detection in concrete." In *Nondestructive testing*, edited by H. S. Lew, 1–20. SP-112. Farmington Hills, MI: American Concrete Institute.

Sansalone, M., and R. Poston. 1992. "Detecting cracks in the beams and columns of a post-tensioned parking garage structure using the impact-echo method." In *Proc., Nondestructive Evaluation of Civil Structures and Materials*, edited by B. A. Suprenant, J. L. Noland, and M. P. Schuller, 129–143. Boulder, CO: University of Colorado.

Shahawy, M. A., and M. Issa. 1992. "Load testing of transversely prestressed double tee bridges." *PCI J.* 37 (2): 86–99.

She, A. C., K. D. Hjelmstad, and T. S. Huang. 1992. "Structural damage detection using stereo camera measurements." In *Proc., Nondestructive Evaluation of Civil Structures and Materials*, edited by B. A. Suprenant, J. L. Noland, and M. P. Schuller, 345–356. Boulder, CO: University of Colorado.

St. John, D. A. 1990. "The use of large-area thin sectioning in the petrographic examination of concrete." In *Petrography applied to concrete and concrete aggregates*, edited by B. Erlin and D. Stark, 55–70. ASTM STP 1061. West Conshohocken, PA: ASTM.

Sturrup, V. R., F. J. Vecchio, and H. Caratin. 1984. "Pulse velocity as a measure of concrete compressive strength." In *In situ/nondestructive testing of concrete*, edited by V. M. Malhotra, 201–227. SP-82. Farmington Hills, MI: American Concrete Institute.

Tamura, H., and M. Yoshida. 1984. "Nondestructive method of detecting steel corrosion in concrete." In *In situ/nondestructive testing of concrete*, edited by V. M. Malhotra, 689–702. SP-82. Farmington Hills, MI: American Concrete Institute.

Taylor, M. A. 1988. "Evaluation of concrete constituents using photon radiation." In *Nondestructive testing*, edited by H. S. Lew, 41–62. SP-112. Farmington Hills, MI: ACI.

Taylor, M. A. 1990. "Acoustic emission monitoring of concrete quality." In *Proc., Non-Destructive Evaluation of Civil Structures and Materials*, edited by B. A. Suprenant, S. Sture, J. L. Noland, and M. P. Schuller, 87–103. Boulder, CO: University of Colorado.

Teodoru, G. V. 1988. "The use of simultaneous nondestructive tests to predict the compressive strength of concrete." In *Nondestructive testing*, edited by H. S. Lew, 137–152. SP-112. Farmington Hills, MI: ACI.

Turkstra, C. J., P. Zoltanetzky, Jr., H. P. Lim, and C. Gordon. 1988. "A statistical study of the correlation between field penetration strength and field cylinder strength." In *Nondestructive testing*, edited by H. S. Lew, 165–179. SP-112. Farmington Hills, MI: ACI.

Weil, G. J. 1992. "Non-destructive testing of bridge, highway and airport pavements." In *Proc., Nondestructive Evaluation of Civil Structures and Materials*, edited by B. A. Suprenant, J. L. Noland, and M. P. Schuller, 385–394. Boulder, CO: University of Colorado.

Whiting, D. 1991. "Preliminary tests of a new surface air-flow device for rapid in-situ indication of concrete permeability." In *Proc., 70th Annual Meeting*, Paper No. 910071. Washington, DC: Transportation Research Board.

Whiting, D. A., and G. A. Matzkanin. 1992. "Field testing of a nondestructive portable permeability indicator for concrete." In *Proc., Nondestructive Evaluation of Civil Structures and Materials*, edited by B. A. Suprenant, J. L. Noland, and M. P. Schuller, 51–67. Boulder, CO: University of Colorado.

Worthington, W. 1992. "Prestressed concrete pipe inspection and monitoring methods." In *Proc., Nondestructive Evaluation of Civil Structures and Materials*, edited by B. A. Suprenant, J. L. Noland, and M. P. Schuller, 83–97. Boulder, CO: University of Colorado.

Yun, C. H., K. R. Choi, S. Y. Kim, and Y. C. Song. 1988. "Comparative evaluation of nondestructive test methods for in-place strength determination." In *Nondestructive testing*, edited by H. S. Lew, 111–136. SP-112. Farmington Hills, MI: American Concrete Institute.

METALS

Greimann, L., and J. Stecker. 1990. "Maintenance and repair of steel sheet pile structures." *Tech. Rep. REMR-OM-9*. Champaign, IL: US Army Construction Engineering Research Laboratory.

O'Dell, D. A., R. E. Hansen, and S. L. McDowell. 1992. "Ultrasonic testing of fillet welds." In *Proc., Nondestructive Evaluation of Civil Structures and Materials*, edited by B. A. Suprenant, J. L. Noland, and M. P. Schuller, 411–425. Boulder, CO: University of Colorado.

Woodham, D. B. 1992. "Magnetically induced velocity change to measure residual stress." In *Proc., Nondestructive Evaluation of Civil Structures and Materials*, edited by B. A. Suprenant, J. L. Noland, and M. P. Schuller, 297–311. Boulder, CO: University of Colorado.

MASONRY

Berra, M., L. Binda, L. Anti, and A. Fatticcioni. 1992. "Utilization of sonic tests to evaluate damaged and repaired masonries." In *Proc., Nondestructive Evaluation of Civil Structures and Materials*, edited by B. A. Suprenant, J. L. Noland, and M. P. Schuller, 329–338. Boulder, CO: University of Colorado.

Forde, M. C. 1992. "Non-destructive evaluation of masonry bridges." In *Proc., Nondestructive Evaluation of Civil Structures and Materials*, edited by B. A. Suprenant, J. L. Noland, and M. P. Schuller, 17–42. Boulder, CO: University of Colorado.

Kayser, J. R., and E. D. Miller. 1992. "Accuracy in the thickness measurement of corroded bridge steel using an ultrasonic thickness meter." In *Proc., Nondestructive Evaluation of Civil Structures and Materials*, edited by B. A. Suprenant, J. L. Noland, and M. P. Schuller, 241–246. Boulder, CO: University of Colorado.

Kisters, F. H., and F. W. Kearney. 1991. "Evaluation of civil works metal structures." *Tech. Rep. REMR-CS-31*. Champaign, IL: US Army Construction Engineering Research Laboratory.

Maji, A. K., and J. L. Wang. 1992. "Inspection of model steel bridge components with electronic shearography." In *Proc., Nondestructive Evaluation of Civil Structures and Materials*, edited by B. A. Suprenant, J. L. Noland, and M. P. Schuller, 263–277. Boulder, CO: University of Colorado.

Wynn, C. C., W. M. Fletcher, and W. D. Jones. 1992. "Nondestructive evaluation of hollow clay tile walls." In *Proc., Nondestructive Evaluation of Civil Structures and Materials*, edited by B. A. Suprenant, J. L. Noland, and M. P. Schuller, 375–383. Boulder, CO: University of Colorado.

WOOD

Anthony, R. W. 1992. "Structural timber assessment using nondestructive evaluation." In *Proc., Nondestructive Evaluation of Civil Structures and Materials*, edited by B. A. Suprenant, J. L. Noland, and M. P. Schuller, 357–373. Boulder, CO: University of Colorado.

ASTM International. 2009. Standard terminology relating to wood and wood-based products. ASTM D9-09a.

Gibson, R. F. 1985. "Frequency domain testing of materials." In *Proc., 5th Nondestructive Testing of Wood Symp., Pullman, Washington, Conf. and Institutes*, 385–406. Pullman, WA: Washington State University.

Humphrey, P. E., and R. L. Ethington. 1990. "Non-destructive testing research at Oregon State University, Dept. of Forest Products." In *Proc., 7th Int. Nondestructive Testing of Wood Symp., Madison, Wisconsin, Conf. and Institutes*, 125–129. Pullman, WA: Washington State University.

Lemaster, R. L., and A. D. Pugel. 1990. "Measurement of density profiles of wood product materials using acoustic emission." In *Proc., 7th Int. Nondestructive Testing of Wood Symp., Madison, Wisconsin, Conf. and Institutes*, 3–28. Pullman, WA: Washington State University.

McCuen, R. H., M. S. Aggour, and B. M. Ayyub. 1988. "Spacing for accuracy in ultrasonic testing of bridge timber piles." *J. Struct. Eng.* 114 (2): 2652–2668.

Patton-Mallory, M., and R. C. De Groot. 1990. "Detecting brown-rot decay in southern yellow pine by acousto-ultrasonics." In *Proc., 7th Int. Nondestructive Testing of Wood Symp., Madison, Wisconsin, Conf. and Institutes*, 29–44. Pullman, WA: Washington State University.

Pellerin, R. F., R. C. DeGroot, and G. R. Esenther. 1985. "Nondestructive stress wave measurements of decay and termite attack in experimental wood units." In *Proc., 5th Nondestructive Testing of Wood Symp., Pullman, Washington, Conf. and Institutes*, 319–352. Pullman, WA: Washington State University.

Portala, J.-F., and J. Ciccotelli. 1990. "NDT techniques for evaluating wood characteristics." In *Proc., 7th Int. Nondestructive Testing of Wood Symp., Madison, Wisconsin, Conferences and Institutes*, 98–124. Pullman, WA: Washington State University.

Ross, R. J. 1985. "Stress wave propagation in wood products." In *Proc., 5th Nondestructive Testing of Wood Symp., Pullman, Washington, Conf. and Institutes*, 291–318. Pullman, WA: Washington State University.

Ross, R. J. 1992. "Nondestructive testing of wood." In *Proc., Nondestructive Evaluation of Civil Structures and Materials*, edited by B. A. Suprenant, J. L. Noland, and M. P. Schuller, 43–49. Boulder, CO: University of Colorado.

GENERAL

Alampalli, S., G. Fu, and I. A. Aziz. 1992. "Nondestructive evaluation of highway bridges by dynamic monitoring." In *Proc., Nondestructive Evaluation of Civil Structures and Materials*, edited by B. A. Suprenant, J. L. Noland, and M. P. Schuller, 211–225. Boulder, CO: University of Colorado.

Al-Qadi, I. L. 1992. "The penetration of electromagnetic waves into hot-mix asphalt." In *Proc., Nondestructive Evaluation of Civil Structures and Materials*, edited by B. A. Suprenant, J. L. Noland, and M. P. Schuller, 195–209. Boulder, CO: University of Colorado.

Clark, W. G., Jr., and R. E. Shannon. 1992. "Magnetic tagging for the improved characterization of construction materials." In *Proc., Nondestructive Evaluation of Civil Structures and Materials*, edited by B. A. Suprenant, J. L. Noland, and M. P. Schuller, 113–127. Boulder, CO: University of Colorado.

Hearn, G., and R. Ghia. 1992. "Response-based structural condition monitoring." In *Proc., Nondestructive Evaluation of Civil Structures and Materials*, edited by B. A. Suprenant, J. L. Noland, and M. P. Schuller, 161–169. Boulder, CO: University of Colorado.

Matzkanin, G. A., T. W. Stephens, and A. J. Chalkley. 1992. "Development of a new nondestructive inspection system for bridge structures." In *Proc., Nondestructive Evaluation of Civil Structures and Materials*, edited by B. A. Suprenant, J. L. Noland, and M. P. Schuller. Boulder, CO: University of Colorado.

Mazurek, D. F., S. R. Jordan, D. J. Palazzetti, and G. S. Robertson. 1992. "Damage detectability in bridge structures by vibrational analysis." In

Proc., Nondestructive Evaluation of Civil Structures and Materials, edited by B. A. Suprenant, J. L. Noland, and M. P. Schuller, 181–193. Boulder, CO: University of Colorado.

O'Leary, P. N., D. A. Bagdasarian, and J. T. DeWolf. 1992. "Bridge condition assessment using signatures." In *Proc., Nondestructive Evaluation of Civil Structures and Materials*, edited by B. A. Suprenant, J. L. Noland, and M. P. Schuller, 171–179. Boulder, CO: University of Colorado.

Pastor, J. A., and A. Navas. 1992. "Evaluation of earthquake damaged structures using non-destructive testing methods." In *Proc., Nondestructive Evaluation of Civil Structures and Materials*, edited by B. A. Suprenant, J. L. Noland, and M. P. Schuller, 313–328. Boulder, CO: University of Colorado.

Platkin, D. E. 1991. "REMR management systems—Coastal/shore protection structures: Condition rating procedures for rubble breakwaters and jetties, initial report." *Tech. Rep. REMR-OM-11*. Champaign, IL: US Army Construction Engineering Research Laboratory.

US Secretary of Transportation. 1991. *Highway bridge replacement and rehabilitation program, tenth report of the secretary of transportation to the United States Congress*. Washington, DC: US Dept. of Transportation, Federal Highway Administration.

APPENDIX B3

EVALUATION PROCEDURES AND EVALUATION OF COMPONENTS AND SYSTEMS—REFERENCES

The following references have been organized by specific material or general category.

CONCRETE

ACI (American Concrete Institute). 1992 (1997). *Considerations for design of concrete structures subjected to fatigue loading*. ACI PRC-215. Farmington Hills, MI: ACI.

ACI. 2013. *Report on nondestructive test methods for evaluation of concrete in structures*. ACI PRC-228. Farmington Hills, MI: ACI.

Bennett, E. W. 1982. "Fatigue tests of spliced reinforcement in concrete beams." In *Fatigue of concrete structures*, edited by S. P. Shah, 177–193. SP-75. Farmington Hills, MI: ACI.

Borgard, B., C. Warren, S. Somayaji, and R. Heidersbach. 1990. "Mechanisms of corrosion of steel in concrete." In *Proc., Symp. on Corrosion Rates of Steel in Concrete*, edited by N. S. Berke, V. Chaker, and D. Whiting, 174–188. ASTM STP 1065. West Conshohocken, PA: ASTM.

Cornet, J., D. Pirtz, M. Polivka, Y. Gau, and A. Shimizu. 1980. "Laboratory testing and monitoring of stray current corrosion of prestressed concrete in seawater." In *Proc., Symp. on Corrosion of Reinforcing Steel in Concrete*, edited by D. E. Tonini and J. M. Gaidis, 17–31. ASTM STP 713. West Conshohocken, PA: ASTM.

Dehghan, A., P. Zhang, E. Ossetchkina, D. Sloan, et al. 2019. "Digital microscopy applied to damage rating index for alkali–silica reaction in concrete." In *Advances in cement analysis and concrete petrography*, edited by D. Cong and D. Broton. West Conshohocken, PA: ASTM.

Deschenes, R., Jr., E. Giannini, T. Drimalas, B. Fournier, et al. 2018. "Effects of moisture, temperature and freezing and thawing on alkali–silica reaction." *ACI Mater. J.* 115 (4): 575–584.

Gerwick, B. C., Jr. 1975. "Practical methods of ensuring durability of prestressed concrete ocean structures." In *Durability of concrete*, 317–324. SP-47. Farmington Hills, MI: ACI.

Geza, T., and B. Magyari. 1982. "Fatigue of reinforcements with pressed sleeve splices." In Vol. 37 of *Proc., Fatigue of Steel and Concrete Structures*, 265–271. IABSE Colloquium. Lausanne, Zurich: International Association for Bridge and Structural Engineering.

Griess, J. C., and D. J. Naus. 1980. "Corrosion of steel tendons used in prestressed concrete pressure vessels." In *Proc., Symp. on Corrosion of Reinforcing Steel in Concrete*, edited by D. E. Tonini and J. M. Gaidis, 32–50. ASTM STP 713. West Conshohocken, PA: ASTM.

Hansen, W. C. 1966. "Concrete aggregates—Chemical reactions." In *Significance of tests and properties of concrete and concrete making materials*, 487–496. ASTM STP 169A. West Conshohocken, PA: ASTM.

Hawkins, N. M., and S. P. Shah. 1982. "American Concrete Institute considerations for fatigue." In Vol. 37 of *IABSE Colloquium Lausanne*, 41–50. IABSE Rep. Zurich: International Association for Bridge and Structural Engineering.

Helgason, T., J. M. Hanson, N. F. Somes, W. G. Corley, et al. 1976. "Fatigue strength of high-yield reinforcing bars." *NCHRP Bull. 164*. Washington, DC: Transportation Research Board, National Research Council.

Hobbs, D. W. 1988. *Alkali–silica reaction in concrete*. London: Thomas Telford.

Hussain, S. E. A. S., Al-Gahtani, and Rasheeduzzafar. 1996. "Chloride threshold for corrosion of reinforcement in concrete." *ACI Mater. J.* 93 (6): 534–538.

Isecke, B. 1982. "Collapse of the Berlin Congress Hall prestressed concrete roof." In *Materials performance*, 36–39. Houston, TX: National Association of Corrosion Engineers.

Kerkhoff, B. 2007. "Diagnosis and control of alkali–aggregate reactions in concrete." In *Concrete technology*. Skokie, IL: Portland Cement Association.

Kordina, K., and U. Quast. 1982. "Fatigue tests on couplings of tendons under service conditions." In Vol. 39 of *IABSE Symp., Washington, DC: Maintenance, Repair and Rehabilitation of Bridges, Final Report, IABSE Rep.*, 155–160. Zurich: International Association for Bridge and Structural Engineering.

Lenschow, R. 1982. "Fatigue of concrete structures." In Vol. 37 of *IABSE Colloquium Lausanne, IABSE Rep.*, 15–24. Zurich: International Association for Bridge and Structural Engineering.

Locke, C. E. 1986. "Corrosion of steel in Portland cement concrete: Fundamental studies." In *Corrosion effect of stray currents and the techniques for evaluating corrosion of rebars in concrete*, edited by V. Chaker, 5–13. ASTM STP 906. West Conshohocken, PA: ASTM.

McCoy, W. J. 1966. "Other materials—Mixing and curing water for concrete." In *Significance of tests and properties of concrete and concrete making materials*, 515–521. ASTM STP 169A. West Conshohocken, PA: ASTM.

Mehta, P. K. 1977. "Effect of cement composition on corrosion of reinforcing steel in concrete." In *Chloride corrosion of steel in concrete*, edited by D. E. Tonini and S. W. Dean, 12–19. ASTM STP 629. West Conshohocken, PA: ASTM.

Mehta, P. K., and M. Polivka. 1975. "Sulfate resistance of expansive cement concretes." In *Durability of concrete*, 367–379. ACI SP-47. Farmington Hills, MI: ACI.

Naaman, A. E. 1982. "Fatigue in partially prestressed concrete beams." In *Fatigue of concrete structures*, edited by S. H. Shah, 25–46. SP-75, Paper SP75-2. Farmington Hills, MI: ACI.

Naaman, A. E. 1989. "Fatigue of reinforcement in partially prestressed beams, structural materials." In *Proc., Sessions Related to Structural Materials at Structures Congress '89*, edited by J. F. Orofino, 377–381. Reston, VA: ASCE.

Nurnberger, V. 1982. "Fatigue resistance of reinforcing steel." In Vol. 37 of *IABSE Colloquium, Lausanne, Fatigue of Steel and Concrete Structures, IABSE Rep.*, 213–220. Zurich: International Association for Bridge and Structural Engineering. [In German.]

Peterson, P. C. 1977. "Concrete bridge deck deterioration in Pennsylvania." In *Chloride corrosion of steel in concrete*, edited by D. E. Tonini and S. W. Dean, 61–68. ASTM STP 629. West Conshohocken, PA: ASTM.

Reading T. J. 1975. "Combating sulfate attack in Corps of Engineers concrete construction." In *Durability of concrete*, 343–366. ACI SP-47. Farmington Hills, MI: ACI.

Rothman, P. S., and R. E. Price. 1986. "Detection and considerations of corrosion problems of prestressed concrete cylinder pipe." In *Corrosion effect of stray currents and the techniques for evaluating corrosion of rebars in concrete*, edited by V. Chaker, 92–107. ASTM STP 906. West Conshohocken, PA: ASTM.

Sibbick, R. G., C. LaFleur, and S. Garrity. 2019. "Using fluorescent microscopy as a tool in the determination of water to cementitious binder ratios in hardened concrete samples." In *Advances in cement analysis and concrete petrography*, edited by D. Cong and D. West. West Conshohocken, PA: ASTM.

Snell, L. M., and R. B. Rutledge. 1990. "Procedures to locate steel in concrete masonry structures." In *Proc., Non-Destructive Evaluation of Civil Structures and Materials*, edited by B. A. Suprenant, S. Sture, J. L. Noland, and M. P. Schuller, 351–360. Boulder, CO: University of Colorado.

Soltani, A., K. Harries, B. Shahrooz, H. Russell, et al. 2012. "Fatigue performance of high-strength reinforcing steel." *J. Bridge Eng.* 17 (3): 454–461.

Stark, D. 2006. "Alkali–silica reactions in concrete." In *Significance of test and properties of concrete and concrete-making materials*, edited by J. Lamond and J. Pielert, 401–409. ASTM STP 169D. West Conshohocken, PA: ASTM.

Struble, L., and S. Diamond. 1986. "Influence of cement alkali distribution on expansion due to alkali–silica reaction." In *Alkalies in concrete*, edited by V. H. Dodson, 31–45. ASTM STP 930. West Conshohocken, PA: ASTM.

Swamy, R. N., and M. M. Al-Asali. 1986. "Influence of alkali–silica reaction on the engineering properties of concrete." In *Alkalies in concrete*, edited by V. H. Dodson, 69–86. ASTM STP 930. West Conshohocken, PA: ASTM.

Tassi, G., and B. Magyari. 1982. "Fatigue of reinforcements with pressed sleeve splices." In Vol. 37 of *IABSE colloquium, Lausanne, fatigue of steel and concrete structures, 265–271. IABSE Reports*. Zurich: International Association for Bridge and Structural Engineering.

Tremper, B. 1966. "Hardened concrete—Corrosion of reinforcing steel." In *Significance of tests and properties of concrete and concrete making materials*, 220–229. ASTM STP 169A. West Conshohocken, PA: ASTM.

US Department of Interior. 1981. *Concrete manual.* 8th ed. Washington, DC: US Dept. of Interior, Water and Power Resources Services.

Verbeck, G. J. 1975. "Mechanisms of corrosion of steel in concrete." In *Corrosion of metals in concrete*, edited by L. Pepper, R. G. Pike, and J. A. Willett, 21–38. ACI SP-49. Farmington Hills, MI: ACI.

METALS

ASTM International. 2019. *Standard practice for ultrasonic examination of steel forgings*. ASTM A388/A388M-19. West Conshohocken, PA: ASTM.

ASTM. 2020. *Standard practice for safeguarding against embrittlement of hot-dip galvanized structural steel products and procedures for detecting embrittlement.* ASTM A143-07. West Conshohocken, PA: ASTM.

ASTM. 2021. *Standard practice for liquid penetrant testing*. ASTM E1417/E1417M-21E1. West Conshohocken, PA: ASTM.

Brown, B. F. 1972. "A preface to the problem of stress corrosion cracking." In *Stress corrosion cracking of metals—A state of the art*, 3–15. ASTM STP 518. West Conshohocken, PA: ASTM.

Fontana, M. G., and N. D. Greene. 1986. *Corrosion engineering*. New York: McGraw-Hill.

Mattsson, E. 1982. "The atmospheric corrosion properties of some common structural metals—A comparative study." In *Materials performance*, 9–19. Houston, TX: National Association of Corrosion Engineers.

Munse, W. H. 1964. *Fatigue of welded steel structures*, edited by L. Grover. New York: Welding Research Council,

Raymond, L., ed. 1988. "Hydrogen embrittlement test methods: Current status and projections." In *Hydrogen embrittlement: Prevention and control*, 10–16. ASTM STP 962. West Conshohocken, PA: ASTM.

Walls, J. C., W. W. Sanders, Jr., and W. H. Munse. 1965. "Fatigue behavior of butt-welded reinforcing bars in reinforced concrete beams." *ACI J.* 62 (2): 169–192.

MASONRY

ASTM International. 2018. *Standard guide for assessment and maintenance of exterior dimension stone masonry walls and facades*. ASTM C1496. West Conshohocken, PA: ASTM.

Bell, G., and B. Gabby. 1999. "In-situ evaluation of compressed brick veneer using the flatjack technique." In *Masonry: Materials, testing, and applications*, edited by J. Brisch, H. Francis, and R. Nelson. West Conshohocken, PA: ASTM.

BIA (Brick Industry Association). 2016. "Introduction to energy performance of brick masonry." *Tech. Notes 4*. Reston, VA: BIA.

BIA. 2020. "Mortars for brickwork." *Tech. Notes 8*. Reston, VA: BIA.

Clayford, T. 1985b. "Durability of brick masonry: A review of the literature." In *Masonry, research, application, and problems*, edited by J. T. Conway and J. C. Grogan, 202–238. ASTM STP 871. West Conshohocken, PA: ASTM.

Cyphers, R. A., A. M. Cyphers, J. M. Knorowski, and A. M. Skertic. 2019. "Evaluation of existing moisture content in brick for hygrothermal models in the rehabilitation of mass masonry walls." In *Masonry 2018*, edited by N. Krogstad and W. McGinley. West Conshohocken, PA: ASTM.

ICRI (International Concrete Repair Institute). 2008. *Guide for the evaluation of masonry façade structures*. IRCI 410.1. St. Paul, MN: ICRI.

Raths, C. H. 1985. "Brick masonry wall nonperformance causes." In *Masonry, research, application, and problems*, edited by J. T. Conway and J. C. Grogan, 182–201. ASTM STP 871. West Conshohocken, PA: ASTM.

Scully, J. R., and P. J. Moran. 1988. "The hydrogen embrittlement susceptibility of ferrous alloys: The influence of strain on hydrogen entry and transport." In *Hydrogen embrittlement: Prevention and control*, edited by L. Raymond, 387–402. ASTM STP 962. West Conshohocken, PA: ASTM.

Sheinker, A. A., and J. D. Wood. 1972. "Stress corrosion cracking of a high strength steel." In *Stress corrosion cracking of metals—A state of the art*, 16–38. ASTM STP 518. West Conshohocken, PA: ASTM.

TMS (The Masonry Society). 2017. *Guide for condition assessment of masonry façades*. TMS 1700-17. Longmont, CO: TMS.

WOOD

APT. 2007. "Basics of wood inspection: Considerations for historic preservation." *APT Bull: J. Preserv. Tech.* 38 (2/3), 1–6.

APT. 2010. "Nondestructive evaluation: Wood." *APT Bull: J. Preserv. Tech.* 41 (4) : 1–6.

ASTM International. 2009. *Standard terminology relating to wood and wood-based products.* ASTM D9-09a.

ASTM. 2020. *Methods of test for moisture content of wood.* ASTM D2016 (withdrawn and superseded by ASTM D4442). West Conshohocken, PA: ASTM.

Emmerich, R. W. 1986. "Decay in wood structures." In *Proc., Structures Congress '86: Evaluation and Upgrading of Wood Structures, Case Studies, Sponsored by the Structural Division,* edited by V. K. A. Gopu, 28–34. New Orleans. Reston, VA: ASCE.

FPL (Forest Products Laboratory). 2015. *Nondestructive evaluation of wood, edited by* R. Ross. FPL-GTR-238. Madison, WI: US Department of Agriculture, Forest Service, FPL.

US Department of Agriculture Forest Service. 2015. *Nondestructive evaluation of wood,* edited by R. Ross. FPL-GTR-238. Madison, WI: USDA Forest Service, Forest Products Laboratory.

GENERAL

ACI (American Concrete Institute). 2023. *Code requirements for load testing of existing concrete structures and commentary.* ACI 437.2-22. Farmington Hills, MI: ACI.

ATC (Applied Technology Council). 2004. *Field manual: Safety evaluation of buildings after windstorms and floods.* ATC-45. Redwood City, CA: ATC.

ATC. 2005. *Post-earthquake safety evaluation of buildings.* ATC 20-1. Redwood City, CA: ATC.

ASTM International. 2015. *Standard terminology of building constructions.* ASTM E631-15. West Conshohocken, PA: ASTM.

APPENDIX B4

EVALUATION PROCEDURES AND EVALUATION OF COMPONENTS AND SYSTEMS—SUPPLEMENTARY REFERENCES

The following references have been organized by specific material or general category.

CONCRETE

Andersen, K. T., and N. Thaulow. 1990. "The study of alkali–silica reactions in concrete by the use of fluorescent thin-sections." In *Petrography applied to concrete and concrete aggregates*, edited by B. Erlin and D. Stark, 71–89. ASTM STP 1061. West Conshohocken, PA: ASTM.

Baker, E. A., K. L. Money, and C. B. Sanborn, 1977. "Marine corrosion behavior of bare and metallic-coated steel reinforcing rods in concrete." In *Chloride corrosion of steel in concrete*, edited by D. E. Tonini and S. W. Dean, 30–50. ASTM STP 629. West Conshohocken, PA: ASTM.

Berke, N. S., D. F. Shen, and K. M. Sundberg. 1990. "Comparison of the polarization resistance technique to the macrocell corrosion technique." In *Proc., Symp. on Corrosion Rates of Steel in Concrete*, edited by N. S. Berke, V. Chaker, and D. Whiting, 38–51. ASTM STP 1065. West Conshohocken, PA: ASTM.

Bloem, D. L. 1966. "Concrete aggregates—Soundness and deleterious substances." In *Significance of tests and properties of concrete and concrete making materials*, 497–512. ASTM STP 169A. West Conshohocken, PA: ASTM.

Cady, P. D. 1977. "Corrosion of reinforcing steel in concrete—A general overview of the problem." In *Chloride corrosion of steel in concrete*, edited by D. E. Tonini and S. W. Dean, 3–11. ASTM STP 629. West Conshohocken, PA: ASTM.

Cook, H. K., and W. J. McCoy. 1977. "Influence of chloride in reinforced concrete." In *Chloride corrosion of steel in concrete*, edited by D. E. Tonini and S. W. Dean, 20–29. ASTM STP 629. West Conshohocken, PA: ASTM.

Corley, W. G., J. M. Hanson, and T. Helgason. 1982. "Background of American design procedure for fatigue of concrete." In *Proc., IABSE Colloquium of Steel and Concrete Structures*.

Cornet, I., and B. Bresler. 1980. "Critique of testing procedures related to measuring the performance of galvanized steel reinforcement in concrete." In *Proc., Symp. on Corrosion of Reinforcing Steel in Concrete*, edited by D. E. Tonini and J. M. Gaidis, 160–195. ASTM STP 713. West Conshohocken, PA: ASTM.

Dillard, J., J. Glanville, T. Osiroff, and R. Weyers. 1991. "Surface characterization of reinforcing steel and the interaction of steel with inhibitors in pore solution." In *Proc., 70th Annual Meeting*, Paper No. 910272. Washington, DC: Transportation Research Board.

Dressman, S., T. Osiroff, J. Dillard, J. Glanville, et al. 1991. "A screening test for rebar corrosion inhibitors." In *Proc., 70th Annual Meeting*, Paper No. 910281. Washington, DC: Transportation Research Board.

Farney, J. A., J. R. Graham, and J. E. Backstrom. 1975. "Influence of hot saline and distilled waters on concrete." In *Durability of concrete*, 325–341. SP-47. Farmington Hills, MI: ACI.

Hansson, C. M., and B. Srensen. 1990. "The threshold concentration of chloride in concrete for the initiation of reinforcement corrosion." In *Proc., Symp. on Corrosion Rates of Steel in Concrete*, edited by N. S. Berke, V. Chaker, and D. Whiting, 3–16. ASTM STP 1065. West Conshohocken, PA: ASTM.

Idorn, G. M., and D. M. Roy. 1986. "Opportunities with alkalies in concrete testing, research, and engineering practice." In *Alkalies in concrete*, edited by V. H. Dodson, 5–15. ASTM STP 930. West Conshohocken, PA: ASTM.

Morf, U. 1989. "Durability of high strength bars and wires in tension." In Vol. 57/1 of *IABSE Symp., Lisbon, Durability of Structures*, 199–204. IABSE Rep. Zurich: International Association for Bridge and Structural Engineering. [In German.]

Powers, T. C. 1975. "Freezing effects in concrete." In *Durability of concrete*, 1–11. SP-47. Farmington Hills, MI: ACI.

Skoulikidis, T., A. Tsakopoulos, and T. Moropoulos. 1986. "Accelerated rebar corrosion when connected to lightening conductors and protection of rebars with needles diodes using atmospheric electricity." In *Corrosion effects of stray currents and the techniques for evaluating corrosion of rebars in concrete*, edited by V. Chaker, 15–291. ASTM STP 906. West Conshohocken, PA: ASTM.

Tilly, G. P., and D. S. Moss. 1982. "Long endurance fatigue of steel reinforcement." In Vol. 37 of *IABSE Colloquium, Lausanne, Fatigue of Steel and Concrete Structures*, 229–238. IABSE Rep. Zurich: International Association for Bridge and Structural Engineering.

Tuthill, L. H. 1966. "Hardened concrete—Resistance to chemical attack." In *Significance of tests and properties of concrete and concrete making materials*, 275–289. ASTM STP 169A. West Conshohocken, PA: ASTM.

METALS

Booth, G. S., and J. G. Wylde. 1990. "Procedural considerations relating to the fatigue testing of steel weldments." In *Fatigue and fracture testing of weldments*, edited by J. M. Potter and H. I. McHenry, 3–15. ASTM STP 1058. West Conshohocken, PA: ASTM.

Burget, W., and J. G. Blauel. 1990. "Fracture toughness of manual metal-arc and submerged arc welded joints in normalized carbon–manganese steels." In *Fatigue and fracture testing of weldments*, edited by J. M. Potter and H. I. McHenry, 272–299. ASTM STP 1058. West Conshohocken, PA: ASTM.

Clark, W. G., Jr., and J. D. Landes. 1976. "An evaluation of rising load k_{iscc} testing." In *Stress corrosion–New approaches*, edited by H. L. Craig Jr., 108–127. ASTM STP 610. West Conshohocken, PA: ASTM.

Dean, S. W., Jr. 1976. "Review of recent studies on the mechanism of stress corrosion cracking in austenitic stainless steels." In *Stress corrosion–New approaches*, edited by H. L. Craig Jr., 308–337. ASTM STP 610. West Conshohocken, PA: ASTM.

Fairchild, D. P. 1990. "Fracture toughness testing of weld heat-affected zones in structural steel." In *Fatigue and fracture testing of weldments*, edited by J. M. Potter and H. McHenry, 117–141. ASTM STP 1058. West Conshohocken, PA: ASTM.

Hartt, W. H. 1990. "Corrosion fatigue testing of steels as applicable to offshore structures." In *Corrosion in natural waters*, edited by C. H. Baloun, 54–69. ASTM STP 1086. West Conshohocken, PA: ASTM.

Link, L. R., 1990. "Fatigue crack growth of weldments." In *Fatigue and fracture testing of weldments*, edited by J. M. Potter and H. I. McHenry, 16–33. ASTM STP 1058. West Conshohocken, PA: ASTM.

Thompson, N. G., and J. A. Beavers. 1990. "Effect of ground-water composition on the electrochemical behavior of carbon steel: A statistical experimental study." In *Corrosion in natural waters*, edited by C. H. Baloun, 101–121. ASTM STP 1086. West Conshohocken, PA: ASTM.

Townsend, H. E., Jr. 1972. "Resistance of high strength structural steel to environmental stress corrosion cracking." In *Stress corrosion cracking of metals—A state of the art*, 155–166. ASTM STP 518. West Conshohocken, PA: ASTM.

MASONRY

Grimm, C. T. 1985. "Corrosion of steel in brick masonry." In *Masonry, research, application, and problems*, edited by J. T. Conway and J. C. Grogan, 67–87. ASTM STP 871. West Conshohocken, PA: ASTM.

Grimm, C. T. 1985. "*Durability of brick masonry: A review of the literature.*" In *Masonry, Research, Application, and Problems*, edited by J. T. Conway and J. C. Grogan, 202–238. ASTM STP 871. West Conshohocken, PA: ASTM.

Grimm, C. T. 1988. "Statistical primer for brick masonry." In *Masonry: Materials, design, construction, & maintenance*, edited by H. A. Harris, 169–192. ASTM STP 992. West Conshohocken, PA: ASTM.

Maurenbrecher, A. H. P., and R. J. Brousseau. 1993. "Review of corrosion resistance of metal components in masonry cladding on buildings." *Internal Rep. No. 640*. Ottawa, ON: National Research Council Canada.

Robinson, G. C., and R. H. Brown. 1988. "Inadequacy of property specifications in ASTM C 270." In *Masonry: Materials, design, construction, and maintenance*, edited by H. A. Harris, 7–17. ASTM STP 992. West Conshohocken, PA: ASTM.

Yi, F. L., and R. L. Carrasquillo. 1985. "A study of the thermal behavior of brick under service conditions in a structure." In *Masonry, research, application, and problems*, edited by J. T. Conway and J. C. Grogan, 101–113. ASTM STP 871. West Conshohocken, PA: ASTM.

WOOD

Cramer, S. M., and J. R. Goodman. 1985. "Predicting tensile strength of lumber." In *Proc., 5th Nondestructive Testing of Wood Symp.*, 525–545. Pullman, WA: Washington State University.

Rabbers, F., and L. Norman. 1986. "Alterations to old wood building, evaluation and upgrading of wood structures, case studies." In *Proc., Structures Congress '86*, edited by V. K. A. Gopu, 96–107. Reston, VA: ASCE.

APPENDIX B5
ORGANIZATION REFERENCES

American Association of State Highway and Transportation Officials (AASHTO)
American Concrete Institute (ACI)
American Institute of Steel Construction (AISC)
American Institute of Timber Construction (AITC)
American National Standards Institute (ANSI)
American Plywood Association (APA)
American Society for Metals (ASM)
ASTM International (ASTM)
American Welding Society (AWS)
America Wood Protection Association (AWPA)
Brick Industry Association (BIA)
Building Officials and Code Administrators International (BOCA)
Construction Engineering Research Laboratory (CERL)
Federal Emergency Management Agency (FEMA)
Federal Highway Administration (FHWA), US Department of Transportation
Forest Products Laboratory (FPL), US Forest Service
International Association for Bridge and Structural Engineering (IABSE)
Institute of Civil Engineers (ICE)
International Code Council (ICC)
International Conference of Building Officials (ICBO)
International Masonry Institute (IMI)
International Union of Testing and Research Laboratories for Materials and Structures (RILEM)
Masonry Institute of America (MIA)
Metals Engineering Institute (now ASM Education)

National Association of Corrosion Engineers (NACE International)
National Forest Products Association (NFPA)
National Institute of Standards and Technology (NIST), US Department of Commerce
National Lumber Manufacturers Association
National Park Service, US Department of Interior
National Particleboard Association (NPA)
National Science Foundation (NSF)
Naval Facilities Engineering Command (NFEC)
Northeast Lumber Manufacturers Association (NeLMA)
Prestressed Concrete Institute (PCI)
Southern Building Code Congress International (SBCC)
Southern Forest Products Association (SFPA)
Southern Pine Inspection Bureau
Strategic Highway Research Program (SHRP)
The Masonry Society (TMS)
Transportation Research Board (TRB)
Truss Plate Institute (TPI)
US Department of Agriculture (USDA)
US Department of Commerce (DoC)
US Department of Housing and Urban Development (HUD)
US Nuclear Regulatory Commission (NRC)
Welding Research Council (WRC)
West Coast Lumber Inspection Bureau

INDEX

Note: Page numbers followed by *f* and *t* indicate figures and tables.

FSC
www.fsc.org
MIX
Papier | Fördert
gute Waldnutzung
FSC® C083411

Kolibri 360 GmbH
Ferdinand-Jühlke-Straße 7
99095 Erfurt, Deutschland
produktsicherheit@kolibri360.de